Stimulated Brillouin Scattering
Fundamentals and Applications

Series in Optics and Optoelectronics

Series Editors: **R G W Brown**, University of Nottingham, UK
E R Pike, Kings College, London, UK

Stimulated Brillouin Scattering

Fundamentals and Applications

M J Damzen
The Blackett Laboratory, Imperial College of Science,
Technology and Medicine, London, UK

V I Vlad, V Babin and A Mocofanescu
Institute of Atomic Physics & University of Bucharest,
Bucharest, Romania

CRC Press
Taylor & Francis Group
Boca Raton London New York

CRC Press is an imprint of the
Taylor & Francis Group, an **informa** business

First published 2003 by IOP Publishing Ltd

Published 2019 by CRC Press
Taylor & Francis Group
6000 Broken Sound Parkway NW, Suite 300
Boca Raton, FL 33487-2742

First issued in paperback 2020

ISBN 13: 978-0-367-57848-0 (pbk)
ISBN 13: 978-0-7503-0870-0 (hbk)

Visit the Taylor & Francis Web site at
http://www.taylorandfrancis.com

and the CRC Press Web site at
http://www.crcpress.com

British Library Cataloguing-in-Publication Data

A catalogue record for this book is available from the British Library.

Library of Congress Cataloging-in-Publication Data are available

Cover Design: Victoria Le Billon

Typeset by Academic + Technical, Bristol

Contents

Preface

Brillouin scattering was discovered in 1922 by Louis Brillouin [1]. It is one of a number of characteristic scattering phenomena that occur when light interacts with solid, liquid or gaseous media and corresponds to the scattering of light from thermally-induced acoustical waves (propagating pressure/density waves) present in media at all temperatures. At normal light levels the amount of scattering is small. The characteristics of the scattering can offer interesting information about the properties of the medium (temperature, pressure) and form the basis for remote fibre sensor devices. With intense coherent laser light, the rate of scattering can become so great that the acoustic wave amplitudes increase and the scattered light takes on an exponential growth [2–8]. This regime corresponds to the phenomenon known as stimulated Brillouin scattering (SBS).

With the invention of the ruby laser in 1960, it was just a few years before Chaio *et al* [2] first observed SBS. The conversion of incident light into backward scattered light can approach unity, such that the transmission of even transparent media can be strongly reduced. For very high intensities ($I_L > 10^{12}$ W/cm^2) all substances become rapidly ionized, forming hot and dense plasmas, but even in this extreme case, SBS still occurs [9–14]. In the stimulated regime of Brillouin scattering the strong scattering is accompanied by the production of intense acoustic waves at such high frequencies (10^9–10^{11} Hz) that they are termed as hypersound. The scattered light frequency is down-shifted from the incident light by the acoustic frequency.

Stimulated Brillouin scattering is a most interesting phenomenon in nonlinear optics. The field of SBS has led to the birth of the fascinating subject of optical phase conjugation when this phenomenon was brought to the world's attention by Zel'dovich *et al* in 1972 [15]. In this pioneering experiment, a ruby laser beam was aberrated by a distorting phase plate and focused into a methane gas cell generating a strong backscattered Brillouin wave. After a second passage of the scattered light back through the original aberrating path, the wave compensated for the aberrations, returning the beam to its original high spatial quality. The experiment revealed that the scattered beam had a wavefront that was uniquely spatially-correlated to the incident wave in both amplitude and phase, and known as the phase conjugate, or wavefront-reversed, wave. With the growth of optical communications based on optical fibres, SBS again was a focus of attention and investigation. For narrowband radiation, SBS can be a strong limit to the transmission of high signal powers. In recent years, phase conjugation based on SBS in multimode optical fibre has been a new area of work, and SBS for narrowband amplifiers in telecommunications has been considered. SBS has been used to control laser radiation including correction of distortions in laser amplifiers and correction of aberrations in laser oscillators.

The aim of this book is to take the interested reader through the subject of SBS covering basic theory to applications. Starting at an elementary level, the basic physics and mathematical description of the phenomenon is developed. The book proceeds to a survey of SBS materials (gases, liquids, solids) and their properties, and applications that include optical phase conjugation, optical communications and laser resonators with SBS mirrors. Interspersed with the applied material are chapters providing more in-depth mathematics on the SBS process in one-dimension and proceeding to three-dimension aspects including transverse effects. A range of techniques to enhance the SBS process are also detailed including two-cell SBS, SBS with optical feedback and Brillouin-enhanced four-wave mixing.

Some of the chapters are illustrated with the personal achievements of some of the authors. In particular some of the work presented by the authors in this book is done in the frame of a collaboration agreement between Imperial College of Science, Technology and Medicine, The Blackett Laboratory, Optics Section, London and Institute of Atomic Physics, The National Institute of Laser, Plasma and Radiation Physics, Laser Section, Bucharest, in the frame of the collaboration program between The Royal Society and The Romanian Academy. The authors are very indebted to these prestigious scientific institutions for their support in this work. The personal support given by Professor Radu Grigorovici, Member of the Romanian Academy (and a past Vice-President of the Romanian Academy) and by Professor Chris Dainty, the Head of Optics Section in The Blackett Laboratory, Imperial College, London, for this scientific collaboration and common work, since 1991, is recalled with gratitude. Dr Damzen would like to acknowledge support from the UK Research Councils for the work. The Romanian team thanks the Romanian Ministry for Education and Research for the financial support offered in several projects on nonlinear optics, which gave the possibility of a continuous effort in this important scientific field. The scientific discussions with distinguished colleagues in the field such as Prof. H Walther, Prof. H Eichler, Prof. A Friesem, Prof. M Petrov, Prof. T Tschudi and Dr. D Proch were very stimulating and we thank them for their time and knowledge offered to us in these discussions.

References

1 Brillouin L 1922 *Ann. Phys.* **17** 88
2 Chiao R Y, Townes C H and Stoicheff B P 1964 *Phys. Rev. Lett.* **12** 592
3 Boyd R W 1992 *Nonlinear Optics* (Boston: Academic) ch 7–9
4 Gower M and Proch D 1994 *Optical Phase Conjugation* (Berlin: Springer Verlag) chap 2–4, 10 and 12
5 Armstrong J A, Bloembergen N, Ducuing J and Pershan P S 1962 *Phys. Rev.* **127** 1918
6 Fabelinskii L 1975 Stimulated Mandelstam–Brillouin Process in *Quantum Electronics: A Treatise* vol I eds H Rabin and C L Tang (New York: Academic) Part A
7 Yariv A 1975 *Quantum Electronics* 2nd edition (New York: Wiley) p 387
8 Kaiser W and Maier M 1972 Stimulated Rayleigh, Brillouin and Raman spectroscopy in *Laser Handbook* vol 2 ed F T Arecchi p 1077
9 Agarwal R N, Tripathi V K and Agarwal P C 1996 *IEEE Trans. on Plasma Science* **24** 143
10 Young P E, Foord M E, Maximov A V and Rozmus W 1996 *Phys. Rev. Lett.* **77**(7) 1278
11 Schmitt A J and Afeyan B B 1998 *Physics of Plasmas* **5** 503
12 Andreev A A and Sutyagin A N 1997 *Quantum Electronics* **27** 150
13 Labaune C, Baldis H A, Schifano E and Tikhonchuk V T 1997 *Journal de Physique III* **7** 1729
14 Baton S D, Amiranoff F, Malka V, Modena A, Salvati M, Coulaud C, Rousseaux C, Renard N, Mounaix P. and Stenz C 1998 *Phys. Rev. E* **57**, R4895
15 Zel'dovich B Ya, Popovichev V I, Ragulsky V V and Faizullov F S 1972 *Sov. Phys. JETP* **15** 109 (English translation)

Chapter 1

Spontaneous and stimulated scattering of light

When light, or other frequency of the electromagnetic spectrum, travels through matter various scattering processes can occur. The matter may be in the form of solid, liquid or gas, but in each case light is scattered by fluctuations or excitations of the optical properties of the medium. The scattering process removes incident photons of light and produces scatter photons that are generally shifted in direction and frequency from the original light. Well-known examples of scattering are Rayleigh scattering, Brillouin scattering and Raman scattering. Under normal light conditions, the scattering is a random, statistical process with scattering occurring over a wide angular spectrum as indicated in figure 1.1. In this chapter, we concentrate on a discussion of the Brillouin scattering phenomenon, primarily. We will however briefly mention the Raman process which also plays an important role in nonlinear optics and can compete with Brillouin scattering as the dominant scattering mechanism [1.1–1.6].

At its most fundamental level, scattering can be described by a quantum mechanical approach, although in practice the origin of some forms of scattering can be adequately described by classical mechanisms (e.g. Brillouin scattering where phonon energy is less than $k_B T$, where k_B is Boltzmann's constant and T is temperature). Scattering occurs due to the interaction of the (classical) light *wave* with excitation (oscillations) in the medium. In quantum theory, the light can be considered as *photons* (quanta of the electromagnetic field) and the medium excitation as *phonons* (quanta of medium excitation). For very low light levels (low photon density) it is necessary to describe the process using photons and phonons. In practice, the light intensity is high (e.g. laser light) and the medium may have strong excitation—in this high quanta limit it is appropriate to use semi-classical wave theory to describe the interaction. In this book, we shall take the wave approach to describe the light–matter interactions in a quantitative manner. We shall, however, also use the picture of photons and medium phonons as this provides a useful complementary description and further insight into the processes.

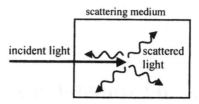

Figure 1.1. Spontaneous scattering of incident light.

Brillouin scattered light originates from light interaction with propagating acoustic waves (or acoustic phonons). Incident photons are annihilated, which together with the creation or annihilation of one phonon gives rise to scattered photons (radiation) at the so-called Stokes or anti-Stokes frequencies, respectively. The Stokes component is downshifted in frequency and the anti-Stokes upshifted. When looking at the frequency spectrum of scattered light in a particular direction (see figure 1.2) these appear as two lines that are close to the incident frequency since the acoustic frequency is much smaller than the optical one, and are called the Brillouin doublet (figure 1.2). On the other hand, when the light is scattered by molecular vibrations, or optical phonons, we speak of Raman scattering. Various frequency shifts, between several hundred and several thousand wavenumbers (cm^{-1}), can occur by Raman scattering and are determined by different vibrational (and rotational) frequencies of the material. This frequency shift can be comparable with the optical frequency (e.g. green light at wavelength $\lambda = 500$ nm has frequency in wavenumbers $= 1/\lambda = 20\,000$ cm^{-1}). Polyatomic molecules or crystals with several atoms in the unit cell frequently exhibit a very rich vibrational and rotational frequency spectrum. In figure 1.2, just a single Raman line is depicted for simplicity. At line centre, there is a Rayleigh scattering component. It has no frequency shift and is due to scattering from non-propagating density fluctuations. As with other forms of scattering

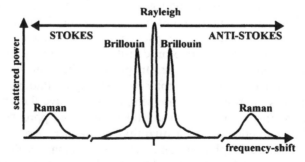

Figure 1.2. The frequency spectrum of spontaneously scattered light. Frequencies are relative to the incident light and shifted to both lower frequency (Stokes shifted) and higher frequency (anti-Stokes shifted).

it is wavelength dependent, being strongest for short wavelengths (e.g. blue light), and it is the mechanism responsible for producing the familiar blueness of the sky during the day.

At low light level the scattering is known as *spontaneous*. In this regime the scattering is caused by the quantum mechanical or thermal excitation of the medium and the amount of scattering is proportional to the incident intensity. On the other hand, at high light intensity the amount of scattering may become strong and material fluctuation may be induced by the presence of the incident light itself. In this regime the light scattering is known as *stimulated*. The stimulated scattering processes are readily observed when the light intensity reaches the range: $10^6 < I_L < 10^9 \, \text{W/cm}^2$. In this intensity range, strong interaction between light fields and matter is observed. For instance, the conversion of incident light into scattered light approaches unity in a number of stimulated scattering processes. As a result, the transmission of a transparent medium is strongly reduced in many cases. By contrast, spontaneous scattering is known to be exceedingly weak (e.g. fractional scattering $\sim 10^{-5}$), so that it has little effect on the light transmission.

As we will see in a later section, an exponential amplification of the scattered light can be expected in the stimulated scattering regime

$$I_S(z) = I_S(0) \exp(g_B I_L l) \qquad (1.1)$$

where $I_S(z)$ is the scattered light intensity at position z in the medium, $I_S(0)$ is the initial level of scattering, g_B denotes the characteristic exponential gain factor of the scattering process, I_L is the intensity of the incident light beam, and l is the length over which amplification takes place.

For very high intensities ($I_L > 10^{12} \, \text{W/cm}^2$), solids, liquids and gases become rapidly ionized, forming hot plasma. Interestingly, stimulated light scattering is obtained even in this extreme plasma regime.

1.1   Spontaneous scattering process

Consider an electromagnetic field (e.g. a light wave) incident on a scattering medium and the case of Stokes scattering where the scattered wave is downshifted in frequency as shown in figure 1.3. Photons of incident light field (E_L) are annihilated, with the creation of photons of scattered light (E_S) and phonons of medium excitation (Q).

Three main quantities characterize the individual scattering process [1.6], as follows.

The frequency shift, $\omega_Q = \omega_L - \omega_S$ (for Stokes scattering). This is determined by the energy ($\hbar\omega$) and momentum ($\hbar\mathbf{K}$) conservations,

$$\omega_L = \omega_S + \omega_Q$$
$$\mathbf{K}_L = \mathbf{K}_S + \mathbf{K}_Q \qquad (1.2)$$

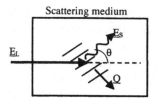

Figure 1.3. A scattering interaction involving an incident photon, scattered photon and medium phonon.

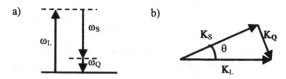

Figure 1.4. Photon–phonon picture of scattering interaction showing relationships between (a) frequencies and (b) wavevectors.

where ω_L, ω_S, ω_Q are the angular frequencies and $\mathbf{K}_L$, $\mathbf{K}_S$, $\mathbf{K}_Q$ are the wavevectors of the incident light, the scattered light and the material excitation (phonon), respectively. Figure 1.4 shows a pictorial representation of the relationships of equation (1.2). The dispersion relationship that relates the wavevectors to frequencies determines the momentum matched frequency shift that characterizes scattering in a given direction θ.

A frequency broadening of the scattered radiation occurs due to the finite lifetime of the excitation, τ_Q. For an exponentially damped matter excitation, the line shape is Lorentzian with the full width at half maximum intensity, $\Delta\nu$, given by

$$\Delta\nu = \frac{\Gamma_Q}{2\pi} = \frac{1}{2\pi\tau_Q} \tag{1.3}$$

where Γ_Q is the angular frequency linewidth which is equal to the reciprocal of the excitation lifetime.

The scattering cross section per steradian $(d\sigma/d\Omega)$ is defined by

$$dP_S/dz = P_L (d\sigma/d\Omega) \Delta\Omega \tag{1.4}$$

where P_S and P_L are the powers of scattered and incident light respectively, and $\Delta\Omega$ is the solid angle. The scattering cross section $(d\sigma/d\Omega)$ is a measure of the strength of scattering in the medium and can be deduced from absolute power measurements of the scattered radiation.

1.2   Brillouin scattering of light

The phenomenon of Brillouin scattering originates from light interaction with a propagating acoustic wave (acoustic phonons). The Brillouin frequency shift ω_B is orders of magnitude smaller than the optical frequency, so to a first approximation $\omega_L \approx \omega_S$ and $|\mathbf{K}_L| \approx |\mathbf{K}_S|$. The wavevector triangle of figure 1.4(b) is an isosceles triangle, and simple calculation shows that the magnitude of the frequency shift ω_B is in this case

$$\omega_B = K_B v = 2\omega_L (n/c) v \sin \theta/2 \qquad (1.5)$$

where c/n is the velocity of light in the medium, v is the velocity of the acoustic phonons and θ is the angle between the wavevectors $\mathbf{K}_L$ and $\mathbf{K}_S$. According to equation (1.5), the value of ω_B is largest for the backward scattered light ($\theta = 180°$). Typical Brillouin shifts are in the range of $\nu_B \approx 0.3$–$6\,\text{GHz}$ (0.1–$2\,\text{cm}^{-1}$) for $\theta = 180°$. Such high acoustic frequencies are known as hypersound. The frequency shift tends to zero at $\theta = 0°$. Since the acoustic lifetime tends to infinity (see below) forward scattering is not observed.

The lifetime of the acoustical phonons, τ_B, is determined by viscous damping mechanisms:

$$\tau_B = \frac{\rho_0}{\eta K_B^2} = \frac{\rho_0}{4\eta\omega_L^2 (n/c)^2 \sin^2(\theta/2)} \qquad (1.6)$$

where ρ_0 is the medium density, η is the viscosity and K_B is the Brillouin wavevector of the acoustic wave. In liquids, values of $\tau_B \approx 10^{-9}\,\text{s}$ are found for $\theta = 180°$, corresponding (from equation 1.3) to a linewidth of $\Delta\nu_B \approx 10^8\,\text{Hz}$ ($3 \times 10^{-3}\,\text{cm}^{-1}$). It is noted that the lifetime is very short at the high acoustic frequencies. The acoustic velocity v is typically $\sim 10^3\,\text{m/s}$ in liquids, so the characteristic propagation length of the damped hypersound ($\nu_B \approx 1\,\text{GHz}$) is $l_B = v\tau_B \approx 1\,\mu\text{m}$.

The scattering cross-section for Brillouin scattering is typically small and has value $(d\sigma/d\Omega) \approx 10^{-6}\,\text{cm}^{-1}\,\text{ster}^{-1}$.

1.3 Raman scattering of light

Brillouin scattering is due to an excitation of the bulk property of the material (e.g. propagating period density fluctuations of an acoustic wave in the medium). The acoustic frequency occurs in a continuum spectrum with the strongest light interaction occurring at simultaneous conservation of energy and momentum in the process. Raman scattering, however, is due to light interaction with resonant modes of a molecular system and the frequency shift is determined by these discrete molecular resonances. The simplest such system is a diatomic molecule (e.g. the nitrogen dimer N_2) as shown in figure 1.5. The strongest Raman scattering is usually associated with vibrational modes of the molecule (vibrational Raman scattering). Rotational modes may also give rise to rotational Raman scattering. The frequency shift of rotational Raman scattering is typically an order of magnitude smaller than that due to the vibrational Raman scattering. The

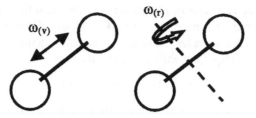

Figure 1.5. Vibrational and rotational modes of oscillation of a diatomic molecule that characterize the Raman scattering process.

molecular vibration lifetime can be deduced from the width of the Raman line (equation 1.3). In liquids, typical values for Raman scattering are: vibrational mode frequency $\nu_R \approx 1000\,\mathrm{cm}^{-1}$ $(3 \times 10^{13}\,\mathrm{Hz})$, linewidth $\Delta\nu_R \approx 5\,\mathrm{cm}^{-1}$ (30 GHz) and the molecular vibration lifetime $\tau_R \approx 10^{-12}\,\mathrm{s}$ (1 ps). The scattering cross-section is $\sim 10^{-7}\,\mathrm{cm}^{-1}\,\mathrm{ster}^{-1}$.

The frequency shift in Raman scattering is a significant fraction of the optical frequency. For radiation at wavelength $\lambda = 500\,\mathrm{nm}$ $(\nu = 2 \times 10^4\,\mathrm{cm}^{-1})$ a Raman shift of $\nu_R = 1000\,\mathrm{cm}^{-1}$ corresponds to a fraction shift $(\nu_R/\nu) = 5\%$ or approximately 25 nm.

For each molecular vibration, two Raman lines can be observed. The Stokes and anti-Stokes lines are connected with transitions from the ground to the first excited state and vice versa. Since, at room temperature, the excited vibrational states are only slightly populated the anti-Stokes intensities are small compared with the Stokes intensities for spontaneous scattering.

As a summary comparison of Brillouin scattering and vibrational Raman scattering, table 1.1 compares typical values of the frequency shift, $\nu = \omega/2\pi$, of the line width of the scattered light, $\Delta\nu$, and of the scattering cross section, $d\sigma/d\Omega$.

Table 1.1

Scattering process	Frequency shift $\nu = \omega/2\pi$ (Hz)	Linewidth $\Delta\nu$ (Hz)	Cross-section $(d\sigma/d\Omega)$ $(\mathrm{cm}^{-1}\,\mathrm{ster}^{-1})$	Gain coefficient g (cm/MW)	Lifetime τ (s)
Brillouin	10^9	10^8	10^{-6}	10^{-2}	10^{-9}
Raman	10^{13}	10^{11}	10^{-7}	5×10^{-3}	10^{-12}

1.4 Stimulated scattering process

Stimulated light scattering [1.7, 1.8] differs from spontaneous scattering in a number of ways:

- it is observed at high light intensity above a certain threshold intensity [1.9];
- it occurs for quasi-monochromatic radiation with a narrow spectral width;
- the scattered spectrum shows distinct line narrowing;
- there is no anti-Stokes component involved in the scattering.

The stimulated Brillouin scattering (SBS) and stimulated Raman scattering (SRS) processes are clearly enhanced by the use of intense radiation. The scattering is also a coherent process as required by energy and momentum conservation conditions and hence requires sufficient temporal and spatial coherence of the radiation source (i.e. lasers). The spontaneous scattering process can be described by linear optics. At high incident and hence scattered light intensity the interaction of these radiations can induce nonlinear enhancement of the medium excitation leading to enhancement in the scattering. Above a critical incident intensity, this enhancement process leads to a positive feedback, and the regime of stimulated light scattering is produced and characterized by exponential amplification of the scattered radiation:

$$I_S(\text{output}) = I_S(\text{input}) \exp(g_B(\nu)I_L l). \tag{1.7}$$

The amplification $G = \exp(g_B(\nu)I_L l)$ is determined by the gain coefficient of the specific scattering process, $g_B(\nu)$, the incident light intensity, I_L, and the medium length (the interaction length) l. It is seen that there is a very strong (exponential) dependence of the scattering with incident intensity. In this way, increase in incident intensity by a small factor can lead to changes in the scattered light by orders of magnitude. In addition, the scattering is most intense at the frequency at gain centre of the stimulated scattering process $g_B(0)$. As a result of the exponential dependence of the light amplification on the gain factor the stimulated scattering is expected to give a narrower linewidth than the spontaneous scattering process.

In SBS the dominant direction of scattering is in the backward direction ($\theta = 180°$). This occurs for several reasons:

(a) The maximum gain length is usually along the axis of the beam and corresponds to the length of the interaction medium (l). For other angles (e.g. 90°) the interaction length is limited by the diameter (d) of the incident beam ($d \ll l$, usually).

(b) The acoustic response time τ_B is a minimum at $\theta = 180°$ (see equation 1.6), so the fastest growth and onset of scattering will be in the backward direction. This is especially important when using short pulse interactions.

(c) When the interaction involves light with a complex wavefront and spatial structure, higher growth rate occurs for a scattered mode that is spatially correlated with the incident beam. This correlation can only occur over a cumulative interaction distance for backward (or forward) scattering. It is this spatial correlation of the scattered

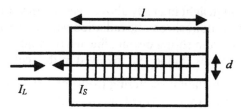

Figure 1.6. Stimulated back-scattering ($\theta = 180°$) is the dominant mechanism for SBS.

radiation that leads to the interesting and important phenomenon known as phase conjugation or also known as wavefront reversal.

In most experimental cases, the source of initial scattered intensity $I_S(\text{input})$ is from the weak spontaneous scattering. For the backscattering geometry of figure 1.6, the spontaneous scattering occurring at the back of the scattering interaction region forms the effective input seeding which is proportional to the laser intensity with typical room temperature value of $I_s(l) \approx I_L(l) \exp(-30)$. When the exponential gain $G \sim \exp(30) \approx 10^{13}$, the stimulated amplification process brings the initial scattering up to a value $I_s(0)$ that is comparable with the incident laser intensity. It is common parlance to talk of a threshold intensity for the onset of SBS due to the strong exponential dependence of growth on incident intensity. This threshold gain factor is commonly taken as $G = \exp(30)$, corresponding to a threshold intensity

$$I_{\text{L,th}} = \frac{30}{g_B l}. \tag{1.8}$$

At this intensity, high conversion of the laser to scattered radiation will be expected to take place. Indeed, the growth equation (1.7) predicts 100% conversion efficiency. This is not correct as the laser intensity is not constant over the interaction length and a fuller system of equations, allowing for laser depletion, needs to be formulated. A useful alternative definition is sometimes used for threshold when $G = \exp(25)$, corresponding to the conversion of about 1% of laser into Stokes radiation (and the approximation of no laser depletion can be assumed). In experimental work, this forms a useful 'observational' definition of when a readily measurable fraction of scattering (relative to incident light) is produced. In reality, the stimulated process still occurs at much lower amplifications and incident intensities, so it must be remembered that the standard definition of threshold is only an 'observational threshold'. This definition is most useful for experimental work, allowing specification of the required intensity level at which strong SBS occurs and when it must be considered.

When operating at intensities much higher than threshold, high conversion efficiencies of laser into scattered light of greater than 90% have been measured. It is this high conversion that makes the stimulated scattering process of such high interest. Depending on application, it can be an advantageous process or have a detrimental effect. The advantage, and potential application, of stimulated scattering is that it can form an efficient nonlinear optical 'mirror' with some unique reflectivity properties, as described in further chapters of this book. Alternatively, at high light intensity, a transparent medium can become almost non-transmitting and this can be a severe problem in some applications, e.g. light transmission through optical fibres in optical communications systems.

1.5 Wave equation and nonlinear polarization

The previous section describes the process of stimulated scattering at a qualitative level. To obtain the detailed behaviour of the process it is necessary to consider the propagation of light through a material and incorporate the material response to the light interaction.

The interaction of the light field with the scattering medium is described by Maxwell's equations [1.11]:

$$\nabla \times \mathbf{E} = -\frac{\partial \mathbf{B}}{\partial t}, \qquad \nabla \times \mathbf{B} = \varepsilon\mu\frac{\partial \mathbf{E}}{\partial t} \qquad (1.9)$$

which combined give the nonlinear wave equation

$$\nabla^2\mathbf{E} - \frac{n^2}{c^2}\frac{\partial^2 \mathbf{E}}{\partial t^2} - \frac{\alpha n}{c}\frac{\partial \mathbf{E}}{\partial t} = \mu_0\frac{\partial^2 \mathbf{P}^{\mathrm{NL}}}{\partial t^2} \qquad (1.10)$$

where $\mathbf{E}$ and $\mathbf{B}$ are the electric and magnetic field vectors, respectively, n is the (linear) refractive index of the medium, α is the (power) absorption coefficient and the operator $\nabla^2 = \partial^2/\partial x^2 + \partial^2/\partial y^2 + \partial^2/\partial z^2$. The refractive index and absorption arise from the linear (low intensity) component of the medium polarization and $\mathbf{P}^{\mathrm{NL}}$ is the nonlinear polarization.

To simplify the wave equation (1.10), some common assumptions are introduced: the electric field consists of a sum of waves of different frequencies (ω_j) with slowly varying amplitudes and phases (E_j), linear polarization and light propagation in the $\pm z$ direction ($\bar{K}_j = \pm K_j\hat{z}$) are assumed. Thus, the electric field and the nonlinear polarization can be written

$$E(\mathbf{r}, t) = \frac{1}{2}\sum_{j=1}^{N}\{E_j\exp[\mathrm{i}(+K_jz - \omega_jt)] + E_j^*\exp[-\mathrm{i}(+K_jz - \omega_jt)]\} \qquad (1.11)$$

$$P^{\mathrm{NL}}(\mathbf{r}, t) = \frac{1}{2}\sum_{j=1}^{N}\{P_j^{\mathrm{NL}}\exp(-\mathrm{i}\omega_jt) + (P_j^{\mathrm{NL}})^*\exp(\mathrm{i}\omega_jt)\}. \qquad (1.12)$$

The slowly varying envelope approximation, which assumes the amplitude varies slowly compared with an optical wavelength or optical period, allows neglect of second derivatives with respect to space propagation direction (z) and time (t), i.e. $\partial^2 E_j/\partial z^2 \ll K_j(\partial E_j/\partial z)$ and $\partial^2 E_j/\partial t^2 \ll \omega_j(\partial E_j/\partial t)$, the wave equation becomes

$$\mp \frac{i}{2k_j}\nabla_T^2 E_j + \frac{n_j}{c}\frac{\partial E_j}{\partial t} \pm \frac{\partial E_j}{\partial z} + \frac{1}{2}\alpha E_j = \frac{i\omega_j}{2\varepsilon_0 cn_j}P_j^{\text{NL}}\exp(\mp iK_j z) \qquad (1.13)$$

where $\nabla_T^2 = (\partial^2/\partial x^2) + (\partial^2/\partial y^2)$ accounts for transverse diffraction of the wave and the upper signs correspond to propagation in the $+z$ direction and lower signs for the $-z$ direction.

For conventional light fields, the polarization is proportional to the electric field and the proportionality constant, the susceptibility, χ_{ij}, is a material constant and $P^{\text{NL}} = 0$. At high electric fields, χ_{ij} is no longer field independent and one can write the polarization as a power series with higher order susceptibilities $\chi^{(n)}$:

$$P_i = \varepsilon_0\left[\sum_j \chi_{ij}^{(1)}E_j + \sum_{j,k}\chi_{ijk}^{(2)}E_jE_k + \sum_{j,k,l}\chi_{ijkl}^{(3)}E_jE_kE_l + \cdots\right]. \qquad (1.14)$$

The third term in equation (1.14) is connected to the stimulated scattering processes and is responsible for the P^{NL} term on the right-hand side of equation (1.13).

1.6   Theoretical formulation of stimulated Brillouin scattering (SBS)

The nonlinear polarization is related to the density ρ and temperature T by

$$P^{\text{NL}} = \left[\left(\frac{\partial\varepsilon}{\partial\rho}\right)_T\Delta\rho + \left(\frac{\partial\varepsilon}{\partial T}\right)_\rho\Delta T\right]E \qquad (1.15)$$

where $\Delta\rho$ and ΔT are the intensity dependent modifications of the density and temperature amplitudes. The interaction of the laser and Stokes field establish an interference that modulates the medium density and drives an acoustic wave. Two mechanisms are responsible for this. The first term in (1.15) is caused by an induced change in density by electrostriction driven by the electric field interference of laser and Stokes field. This is the major SBS mechanism. The second term in (1.15) occurs in absorbing media where the interference of the fields leads to a temperature wave, which in turn induces density changes and drives an acoustic wave. This latter mechanism is known as stimulated temperature Brillouin scattering (STBS) and only occurs in absorbing media.

In this section, we shall make a number of simplifying approximations to establish the main features of SBS and reintroduce the fuller formulation at a later stage.

(a) We assume no stimulated temperature Brillouin scattering (STBS) by neglecting all intensity-dependent temperature effects ($\Delta T = 0$);
(b) We assume plane waves and neglect transverse diffraction ($\nabla_T = 0$).

Taking small changes in medium density $\Delta\rho = \rho' - \rho_0$ caused by the presence of optical fields, where ρ_0 is the average density, we write fields and density wave in the form

$$E = \tfrac{1}{2}\{E_L \exp[i(K_L z - \omega_L t)] + E_L^* \exp[-i(K_L z - \omega_L t)]$$
$$+ E_S \exp[i(-K_S z - \omega_S t)] + E_S^* \exp[-i(-K_S z - \omega_S t)]\} \quad (1.16)$$

$$\Delta\rho = \tfrac{1}{2}\{\rho \exp[i(Kz - \omega t)] + \text{c.c.}\} \quad (1.17)$$

where the medium is taken to be driven at the difference frequency, $\omega = \omega_L - \omega_S$, of the laser and Stokes fields, and K_L, K_S and K are magnitudes of wavevectors with $K = K_L + K_S \approx 2K_L$. The laser field is travelling in the $+z$ direction, Stokes field in the $-z$ direction (backscattering) and the density (acoustic) wave is travelling in the $+z$ direction. For generality, we allow for a small frequency detuning from resonance $\Delta\omega = \omega - \omega_B$.

The form of equation (1.17) is expected due to the electrostrictive force on the medium that depends on the square of the electric field E^2. This causes the density to increase in the regions of high field strength. The medium is unable to respond directly to the electric field at the optical frequency but can respond to the difference in frequency between two optical fields that have a beat (difference) frequency $\omega = \omega_L - \omega_S$ that propagates with a speed $(\omega_L - \omega_S)/(K_L + K_S)$. If the electrostrictive beat modulation propagates at the speed of sound in the medium the driving force stays in phase with the generated acoustic wave. This is phase matching condition or momentum matching condition that corresponds to resonance.

From equation (1.13), one can immediately obtain the following field equations

$$\frac{\partial E_L}{\partial z} + \frac{n}{c}\frac{\partial E_L}{\partial t} + \frac{1}{2}\alpha E_L = \frac{i\omega_L}{4cn_L}\frac{\gamma_e}{\rho_0} E_S \rho \quad (1.18)$$

$$-\frac{\partial E_S}{\partial z} + \frac{n}{c}\frac{\partial E_S}{\partial t} + \frac{1}{2}\alpha E_S = \frac{i\omega_S}{4cn_S}\frac{\gamma_e}{\rho_0} E_L \rho^* \quad (1.19)$$

where $\gamma_e = \rho_0(\partial\varepsilon/\partial\rho)_T$ is the electrostriction coefficient.

In order to obtain the amplitude of the nonlinear perturbation induced in the medium density (ρ) the Navier–Stokes equation with an electrostrictive force is combined with the equation of continuity to give a material equation:

$$\frac{\partial^2\rho}{\partial t^2} - 2i\omega\frac{\partial\rho}{\partial t} - (\omega^2 - \omega_B^2 + i\omega\Gamma_B)\rho = \frac{\gamma_e\varepsilon_0 K_B^2}{2} E_L E_S^*. \quad (1.20)$$

In this equation the spatial derivatives of the acoustic field have been neglected as the acoustic wave does not propagate significantly on the timescale of the scattering ($v \ll c/n$). This equation has the form of a forced damped harmonic oscillator where the forcing term on the right-hand side is due to electrostriction that causes an increase in density in regions of high electric field.

We can use the approximation $\partial^2 \rho / \partial t^2 \ll 2\omega(\partial \rho / \partial t)$, which is valid if the acoustic wave amplitude growth is slow compared with the acoustic frequency. This condition holds in most practical cases, but breaks down if trying to drive SBS with very short pulses—acoustic period ~ 1 ns. We also consider the case of small detuning from resonance in which case $(\omega^2 - \omega_B^2) = (\omega + \omega_B)(\omega - \omega_B) \approx 2\omega_B \, \Delta\omega$ and we obtain the acoustic wave equation

$$\frac{\partial \rho}{\partial t} + \left(-\mathrm{i}\,\Delta\omega + \frac{\Gamma_B}{2} \right)\rho = \frac{\mathrm{i}\gamma_e \varepsilon_0 K_B}{4v} E_L E_S^*. \tag{1.21}$$

Equations (1.18), (1.19) and (1.21) describe the SBS process, in time and space, on the assumption of plane–wave interaction. We consider a range of solutions to this set of equations under different approximations in chapter 3. It is instructive at this point to seek the steady-state set of equations when the time derivatives are set to zero. In this case, we can derive a simple expression for the density wave:

$$\rho = \frac{\mathrm{i}\gamma_e \varepsilon_0 K_B}{4v} \frac{1}{(1 - 2\mathrm{i}\,\Delta\omega/\Gamma_B)} E_L E_S^*. \tag{1.22}$$

Inserting ρ into the optical field equations (1.18) and (1.19) and converting fields to intensities using the relation $I_j = \varepsilon_0 cn|E_j|^2/2$, we obtain the resulting equations describing the pump intensity, I_L, and the Stokes intensity, I_S:

$$\frac{\mathrm{d}I_L}{\mathrm{d}z} = -g_B(\Delta\omega)I_L I_S \tag{1.23a}$$

$$\frac{\mathrm{d}I_S}{\mathrm{d}z} = -g_B(\Delta\omega)I_L I_S \tag{1.23b}$$

where $g_B(\Delta\omega)$ is the steady-state Brillouin gain coefficient given by

$$g_B(\Delta\omega) = g_B(0)\frac{1}{1 + (2\Delta\omega/\Gamma_B)^2} \tag{1.24a}$$

$$g_B(0) = \frac{\omega_s^2 (\gamma_e)^2}{c^3 nv\rho_0 \Gamma_B} \tag{1.24b}$$

where $g_B(0) = g_B$ is the maximum gain coefficient at resonance, and we have taken $\omega_L = \omega_s$ as a good approximation since the acoustic frequency shift $\sim 10^{-5}$ times the optical frequency. The gain coefficient given by (1.24a) has a Lorentzian angular frequency spectrum with full-width half-maximum of Γ_B. Equations (1.23) show that there is a transfer of energy between the laser and Stokes wave where the laser intensity experiences depletion and the Stokes intensity experiences growth (in the $-z$ direction). The solution of equation (1.23b), with neglect of laser pump depletion, shows that the Stokes intensity grows exponentially with distance as it propagates in the $-z$ direction starting from an initial value at $z = l$:

$$I_S(z) = I_S(l) \exp[g_B(\Delta\omega)I_L(0)(l-z)]. \tag{1.25}$$

Equation (1.21) for the acoustic wave density shows that at resonance and with no optical fields the density wave amplitude decays exponentially with time from an initial value at $t = 0$:

$$\rho(t) = \rho(0) \exp -(\Gamma_B/2)t. \tag{1.26}$$

The intensity of the acoustic wave is proportional to the square of the density amplitude so the acoustic intensity decays with a rate Γ_B, or with a characteristic exponential decay time τ_B given by

$$\tau_B = \frac{1}{\Gamma_B}. \tag{1.27}$$

For cases where the scattered wave is initiated from noise, the process will maximize at resonance. The set of field equations describing the resonant case under steady-state conditions can be readily expanded to incorporate non-plane waves with transverse diffraction and include medium absorption using equation (1.13) and leading to

$$\frac{i}{2K_L}\nabla_T^2 E_L + \frac{\partial E_L}{\partial z} + \frac{n}{c}\frac{\partial E_L}{\partial t} + \frac{1}{2}\alpha E_L = -\frac{g_B}{2}I_S E_L \tag{1.28}$$

$$\frac{i}{2K_S}\nabla_T^2 E_S - \frac{\partial E_S}{\partial z} + \frac{n}{c}\frac{\partial E_S}{\partial t} + \frac{1}{2}\alpha E_S = \frac{g_B}{2}I_L E_S. \tag{1.29}$$

The above pair of equations presents a generalized steady-state set of equations describing the propagation and interaction of laser and Stokes fields. The more generalized case is the fully transient case with the temporal evolution of the acoustic field, and at resonance is given by the following set of three equations:

$$-\frac{i}{2k_L}\nabla_T^2 E_L + \frac{n_L}{c}\frac{\partial E_L}{\partial t} + \frac{\partial E_L}{\partial z} + \frac{1}{2}\alpha_L E_L = \frac{i\omega_L}{4cn}\frac{\gamma_e}{\rho_0}E_S\rho \tag{1.30a}$$

$$+\frac{i}{2k_S}\nabla_T^2 E_S + \frac{n_S}{c}\frac{\partial E_S}{\partial t} - \frac{\partial E_S}{\partial z} + \frac{1}{2}\alpha_S E_S = \frac{i\omega_S}{4cn}\frac{\gamma_e}{\rho_0}E_L\rho^* \tag{1.30b}$$

$$\frac{\partial\rho}{\partial t} + \frac{\Gamma_B}{2}\rho = \frac{i\gamma_e\varepsilon_0 K_B}{4v}E_L E_S^*. \tag{1.30c}$$

This set of equations forms the general set which will be used in subsequent chapters under various sets of approximation.

1.7 Stimulated temperature Brillouin scattering

In addition to the Brillouin process of non-absorbing media, a contribution to the acoustic wave process due to light absorption is found. This process is called stimulated thermal Brillouin scattering (STBS). In STBS, the light absorption generates temperature fluctuations, which in turn produce density fluctuations; this is a secondary source of acoustic wave generation which will couple to the light waves via the change of the index of refraction with density.

It is convenient for the calculation to choose the density, ρ, and the temperature, T, as independent variables. The analysis follows the same procedure as before but with a new temperature variable:

$$\bar{T} = T - T_0, \qquad \bar{T} \ll T_0, \qquad \bar{T} = \tfrac{1}{2}\{T \exp[i(Kz - \omega t)] + \text{c.c.}\}. \quad (1.31)$$

We can derive a new equation for the scattered field (and a similar one exists for the laser field)

$$-\frac{\partial E_S}{\partial z} + \frac{n}{c}\frac{\partial E_S}{\partial t} + \frac{1}{2}\alpha E_S = \frac{i\omega_S}{4cn}\left[\frac{\gamma_e}{\rho_0}E_L\rho^* + \left(\frac{\partial \varepsilon}{\partial T}\right)_\rho E_L T^*\right]. \quad (1.32)$$

Incorporating the combined effects of electrostriction and absorption into the Navier–Stokes equation leads to a gain coefficient with a contribution from SBS and STBS given by

$$g_B(\Delta\omega) = g_B^e(0)\frac{1}{1 + (2\Delta\omega/\Gamma_B)^2} + g_B^a(0)\frac{4\Delta\omega/\Gamma_B}{1 + (2\Delta\omega/\Gamma_B)^2}. \quad (1.33)$$

The electrostrictive term is the same as before and the absorptive term has a value of

$$g_B^a(0) = \frac{\omega_s^2 \gamma^e \gamma^a}{2c^3 n v \rho_0 \Gamma_B} \quad (1.34)$$

where $\gamma^a = \alpha(vc^2\beta_T/C_p\omega_s)$ is called the absorptive coupling constant, β_T is the thermal expansion coefficient and C_p is the specific heat at constant pressure. The absorptive contribution to the gain factor has a maximum that occurs at $\Delta\omega = \Gamma_B/2$. It is zero at the exact resonance frequency for the pure SBS case and can add to the electrostrictive term for $\Delta\omega > 0$ (Stokes side), or subtract for $\Delta\omega < 0$ (anti-Stokes side), in which case g_B may be negative if the absorption is large enough. It is clear that STBS leads to a maximum growth that will occur at a detuning from the resonance frequency of the pure electrostrictive SBS process.

1.8 Comparison of SRS and SBS

As previously mentioned, stimulated Raman scattering (SRS) and stimulated Brillouin scattering (SBS) are well-known inelastic scattering processes resulting from the interaction of light with matter. An input laser wave (pump) at frequency ω_L excites an internal vibration of the medium with frequency (ω_Q) which is simultaneously accompanied by the production of a scattered beam, the Stokes wave, at frequency $\omega_S = \omega_L - \omega_Q$. In SRS the internal resonance is a quantized transition due to a molecular vibration often called the optical phonon, while SBS results from interaction with an acoustic wave, sometimes called the acoustic phonon.

These two scattering processes are quite analogous in many features and in several situations can be described by an identical mathematical formalism. Indeed, in the steady-state Brillouin case the mathematical solutions formulated for SBS can be used for SRS. However, some significant physical differences should be noted.

(i) The frequency shift of SRS is typically of the order of $1000\,\text{cm}^{-1}$ (e.g. $\nu_R = 2914\,\text{cm}^{-1}$ for CH_4) whereas the shift due to SBS is considerably smaller and of the order of $0.1\,\text{cm}^{-1}$ (e.g. $\nu_B \approx 0.1\,\text{cm}^{-1}$ for CH_4 with $\lambda_L = 249\,\text{nm}$). The very small frequency shift of the Brillouin process offers the potential for the backscattered pulse to undergo re-amplification in the laser system since the gain bandwidth of all but a few media is very much larger than the Brillouin shift.

(ii) The damping time of SRS is very short and typically τ_R are a few picoseconds while for SBS, τ_B is in the nanosecond region, which often produces transient phenomena into the scattering process of SBS. It is due to the long damping time in SBS that a transient analysis may need to be performed, often requiring a numerical solution of equations. The model predicts phenomena not present in the steady-state models.

(iii) Forward and backward SRS can occur with maximum gain usually in the forward direction. Due to the phase matching requirement in SBS, only backward scattering occurs, with no scattering in the forward direction possible in isotropic media. The production of second- and higher-order Stokes frequencies in the SRS process due to the forward scattering mechanism strongly limits the operation of Raman scattering at high conversions, although the cascade to higher Stokes and anti-Stokes frequencies can have important application. Higher-order scattering effects have less influence on stimulated Brillouin scattering.

(iv) The Raman steady-state gain coefficient (g_R) usually scales with density of scatterer and pump wavelength as $g_R \propto \rho / \lambda_L$ while for SBS the corresponding gain coefficient varies as $g_B \propto \rho^2$. In the case of gases, the quadratic dependence of density upon the SBS gain normally requires the use of moderately high pressures to reach the very high gains characteristic of the Brillouin process.

(v) The Brillouin process, where the Stokes is spontaneously generated from noise, frequently induces a phase conjugated backscattered signal. In Raman scattering the relatively large frequency shift of the Stokes beam limits the situations under which phase-conjugation may occur, and reduces the degree of optical compensation that may be achieved by double-passing aberrated optical components.

References

1.1   Fabelinskii L 1968 *Molecular Scattering of Light* (New York: Plenum)
1.2   Fabelinskii L 1975 Stimulated Mandelstam–Brillouin Process in *Quantum Electronics: A Treatise* ed H Rabin and C L Tang (New York: Academic Press) vol I part A
1.3   Yariv A 1975 *Quantum Electronics* 2nd edition (New York: Wiley) p 387
1.4   Shen Y R 1984 *Principles of Nonlinear Optics* (New York: Wiley)
1.5   Boyd R W 1992 *Nonlinear Optics* (Boston: Academic) ch 7–9
1.6   Kaiser W and Maier M 1972 Stimulated Rayleigh, Brillouin and Raman spectroscopy in *Laser Handbook* vol 2 ed F T Arecchi p 1077
1.7   Hikita T and Oka N 1996 *Japanese J. Appl. Phys.* **35** 4561
1.8   Chiao R Y, Townes C H and Stoicheff B P 1964 *Phys. Rev. Lett.* **12** 592
1.9   Hyun-Su-Kim, Sung-Ho-Kim, Do-Kyeong-Ko, Gwon-Lim, Byung-Heon-Cha and Jongmin-Lee 1999 *Appl. Phys. Lett.* **74** 1358
1.10 Nguyen-Vo N M and Pfeifer S J 1993 *IEEE J. Quantum Electron.* **29** 508
1.11 Landau L D and Lifshitz E M 1960 *Electrodynamics of Continuous Media* (Reading, MA: Addison-Wesley) ch 14
1.12 Cummins H Z and Gammon R W 1966 *J. Chem. Phys.* **44** 2785
1.13 Enns R H and Batra I P 1969 *Phys. Lett.* **28A** 591
1.14 Afshaarvahid S, Devrelis V and Munch J 1998 *Phys. Rev. A* **57** 3961
1.15 Rother W, Meyer H and Kaiser W 1970 *Phys. Lett.* **31A** 245
1.16 Damzen M J, Hutchinson M H R and Schroeder W A 1987 *IEEE J. Quantum Electron.* **QE-23** 328
1.17 Maier M 1968 *Phys. Rev.* **166** 113
1.18 Tang C L 1966 *J. Appl. Phys.* **37** 2945
1.19 Kroll N M 1965 *J. Appl. Phys.* **36** 34
1.20 Kroll N M and Kelley P L 1971 *Phys. Rev. A* **4** 763
1.21 Pohl D and Kaiser W 1970 *Phys. Rev. B* **1** 31

Chapter 2

Materials for SBS

2.1 The choice of SBS materials and SBS properties

SBS occurs in all states of matter: gases, liquids, solids and plasmas. A vast number of SBS media have been used in a wide variety of experiments over the years, such that it is very difficult to cover a complete list and to fully reference the work of so many authors in the field. The aim of this chapter is to provide a good starting point for the range of suitable materials, with cited papers (2.1–2.47) giving more references for the interested reader.

Choice of state of matter and then specific material is very diverse and depends on a number of requirements. Some materials are more suited for a specified laser wavelength due to their absorption properties and for this reason table 2.1 lists a range of SBS materials according to the laser wavelengths that have commonly been used with that material with the ordering from the infrared to the ultraviolet wavelengths. Some key material parameters related to the SBS process are listed in table 2.1.

The important (but not complete) set of parameters include the following:

The Brillouin frequency shift ν_B (defined for backscattering),

$$\nu_B = \omega_B/2\pi = 2n(v/c)\nu_L = 2v/\lambda_L \qquad (2.1)$$

where v and c/n are the velocities of the sound and light in the material respectively, ν_L is the frequency of the light and λ_L is the light wavelength in vacuum.

The Brillouin linewidth $\Delta\nu_B = \Gamma_B/2\pi$, where Γ_B is the inverse of the acoustic lifetime τ_B and given by

$$\Gamma_B = \frac{1}{\tau_B} = (k_B^2/\rho_0)\eta_{\text{eff}} \qquad (2.2)$$

where $k_B = \omega_B/v$ is the acoustic wavevector, ρ_0 is the material density, and η_{eff} is the (effective) viscosity. This relation is incomplete for gases and a fuller description of the acoustic decay mechanisms will be described later in this chapter.

17

Table 2.1. Selected materials for SBS.

Material	Laser wavelength (pump), λ_L (nm)	Brillouin frequency, ν_B (GHz)	Brillouin linewidth, $\Delta\nu_B$ ($= \Gamma_B/2\pi$) (MHz)	Brillouin-gain coefficient g_B (cm/GW)	Ref.
Xenon gas (40–60 atm)	CO_2, 10600				33
Xenon gas (40–60 atm)	HF, 2900	0.1	<1		5, 6
Sulf hexafluoride (SF_6)	2900				7
Carbon disulphide	Er-doped, 2800	1.42	7		8
Quartz fibre	1550	0.1	60		47
Xenon gas (10–40 atm)	Iodine, 1315p	0.25		120	9
Carbon disulphide (+Xe)	Iodine, 1315p	0.2			10, 11
Methane gas (10^2 atm)	1064	0.864	106–60	8–65	12, 39
Nitrogen gas (135 atm)	1064		45	30	39
SF_6 gas (22 atm)	1064		42	35	39
Xenon gas (40 atm)	1064		33	44	39
Acetone	1060	2.99	119–526	15.8–20	13, 14
Benzene	1060	4.12	228	9.6	13
Carbon disulphide	1060	3.76	50–217	68–130	13, 14
Carbon tetrachloride	1060	2.77	528	3.8–6	13, 14
Silicon tetrachloride	1060			10	9
Titan tetrachloride	1060			20	9
Nitrobenzene	1060	4.26	396	7.2	13
Water	1060	3.70	170	3.8	13
Acetone	694	4.6	180	10.8–20	14, 15
Benzene	694	6.47	289	18	15
Carbon disulphide	694	5.85	66–75	45–130	14, 15
Carbon tetrachloride	694	4.39	650	6	14, 15
Cyclohexane	694	5.55	774	6.8	15
Acetone	694	1.6-4.6	180	10-20	14, 15
n-Hexane	694		212–220	19–26	14
Methanol	694	4.25	250	13	15
Lucite	694		106	17	15
Quartz (x-cut)	694		<40	1.5	14
Schott glass, BK7	694		80	350	14
Schott glass, F2	694		45	2100	14
Schott glass, FK3	694		<40	1800	14
Schott glass, SF6	694		45	2000	14
Water	694	5.69	317	4.8	15
Xenon gas (10 atm)	532	0.65	98.1	1.38	16
Acetone	532	6.0	320	20	17
Carbon disulphide	532	7.7	120	130	17
Calcium fluoride	532	37.2	46	4.11	16
Fused silica	532	32.6	168	2.69	16
Schott glass, BK7	532	35.7	165	2.15	16
Methanol	530	5.6	210	13	17
Borate glass	488	19	100	12.9	18
Halide glass	488	17.64	214	2.8	18
Silicate glass	488	23.4	184	3.9	18
Ethanol, heptane	XeF, 351, p				19
	XeCl, 308, p				20
SF_6 gas	351, p				21, 24

Table 2.1. Continued.

Material	Laser wavelength (pump), λ_L (nm)	Brillouin frequency, ν_B (GHz)	Brillouin linewidth, $\Delta\nu_B$ ($= \Gamma_B/2\pi$) (MHz)	Brillouin gain coefficient g_B (cm/GW)	Ref.
2.2-dyamethilbutane/ n- pentane and n-octane	308				8
Methanol	XeCl, 308, p	1.3		13	22, 23, 24
Cyclohexane	XeCl, 308, p	4		7	24
Quartz-polimer fibre	XeCl, 308, p				25
Ethanol, methanol,	ArF, 193, p	3			19, 26,
isopropyl alcohol,	KrF, 248, p				27, 28
hexane	XeF, 351, p				

The Brillouin-gain intensity coefficient (for on-resonant electrostrictive component) is given by

$$g_B = \left(\frac{\gamma_e^2 k_s^2}{\rho_0 n v c \Gamma_B} \right) \qquad (2.3)$$

where γ_e is the electrostrictive coefficient ($\gamma_e = (n^2 - 1)(n^2 + 2)/3$ in centro-symmetric materials), n is the refractive index and k_s is the wavevector of the Stokes scattered wave.

The Brillouin frequency shift, linewidth (and hence acoustic lifetime) and gain coefficient can be experimentally measured or calculated and some methods to perform this will be described later in this chapter.

A major decision in SBS applications is the selection of a nonlinear material and of the interaction geometry. In general, the use of pressurized gases eliminates complications arising from contamination, cavitation bubbles and thermal convection currents that can occur in liquids. Varying the gas pressure can scale their Brillouin properties, but in general they have relatively long acoustic decay times of many nanoseconds or even tens of nanoseconds. The use of liquids is, however, often selected due to lower SBS thresholds and larger Brillouin linewidths resulting in one order of magnitude shorter acoustic decay times, usually sub-nanosecond. In the first approximation, the Brillouin-gain g_B scales as ρ_0^2 and the acoustic decay time scales as $\rho_0 \lambda^2$. Thus, for the usual Q-switched lasers (with pulses of ~ 10 ns), the gases can show a highly transient behaviour in SBS, while the SBS liquids may be considered in a steady-state regime.

The acoustic decay time scales as the square of the wavelength and the SBS threshold intensity is proportional to the acoustic decay time. This is the reason for poor SBS performances with long wavelength lasers such as CO_2, which display high transiency and low gains. Studies to find improved performance in the mid-infrared spectral region have been undertaken [2.33].

The solid SBS materials are used for all solid-state applications, mostly in guiding geometry (optical fibres), in which long interaction lengths offset their small Brillouin gains. The solid SBS materials are characterized by small linewidths and hence long acoustic decay times of several nanoseconds.

The SBS geometry plays an important role in efficiency and the quality of optical phase conjugation. Much analysis and measurements have shown that better fidelity is obtained by light focusing in a nonlinear material enclosed in a multimode waveguide (or 'light pipe') than in the bulk nonlinear material.

In order to select the optimum material for a particular experimental system, it is important to have a reliable characterization of SBS materials. The criteria for choosing a suitable material depend on a number of factors including laser wavelength and pulse duration, medium absorption, gain coefficient, phonon lifetime and competing nonlinear effects. As previously shown, the main parameters that characterize an SBS material are the acoustic decay time, $\tau_B = (\Gamma_B)^{-1}$, the steady state gain coefficient, g_B, and the acoustic frequency, ν_B. Other subsidiary information of relevance to Brillouin media includes their transparency and thresholds for other nonlinearities, e.g. breakdown, thermally-induced beam defocusing, self-focusing and stimulated Raman scattering. In general, with low-power systems, one will require a high Brillouin gain, while for work with short pulses it is preferable to use materials with short acoustic decay time.

In many applications, the use of pressurized gases is very attractive due to their high transparency throughout the visible region and also into the infrared and ultraviolet regions of the spectrum. By increasing the pressure (and hence density) of the gas, it is also possible to achieve very high gain coefficients, although normally with an increase of the acoustic decay time. The acoustic decay time is probably the least characterized of all the Brillouin parameters. This is partially due to the difficulty of accurately and unambiguously measuring the decay time by experiment. However, the acoustic decay time will determine the degree of transiency and efficiency of these materials for Brillouin scattering, when using relatively short optical pulses. In applications requiring phase conjugate mirrors for the correction of dynamic aberrations, τ_B determines the response time of the scattering to adapt to the temporally varying wavefront of the incoming radiation.

We shall further present a systematic experimental and theoretical determination of the acoustic decay time in several heavy gases over a range of pressures. The Brillouin-gain coefficient g_B and acoustic frequency have also been experimentally determined or calculated for the gases investigated. Similar experiments can be done for the measurement of the principal SBS parameters in liquids. At the end, we shall show a typical setup for the study of Brillouin-gain coefficient and linewidth in standard optical fibres.

In the following, it is shown that the decay of the acoustic wave is generally more complex than given by just the conventional attenuation mechanisms of viscosity and thermal conductivity. In heavy polyatomic gases, decay contributions due to vibrational and rotational relaxation processes are present. Indeed, at the higher density ranges used in the experiments, it is demonstrated, in accordance with theory, that the vibrational relaxation can be the major decay mechanism.

The theoretical results can be extended to other laser wavelengths and more generalized expressions are given to calculate the decay time of other gases.

2.2   Acoustic attenuation mechanisms in polyatomic gases

There are several mechanisms responsible for the attenuation of acoustic waves in a polyatomic gas. The adiabatic density fluctuations comprising the acoustic wave result in a corresponding periodic temperature variation about the mean temperature of the medium. The thermal gradients thereby produced will be attenuated by the viscosity (η) and thermal conductivity (κ) of the medium. This type of decay is present in all media and is described by the sound absorption equation [2.42]:

$$\Gamma_{cl} = \tau_{cl}^{-1} = \frac{k_B^2}{\rho}\left[\frac{4}{3}\eta + \frac{(\gamma-1)\kappa}{c_p}\right] \qquad (2.4)$$

where Γ_{cl} and τ_{cl} are the linewidth and decay time of the acoustic wave resulting from classical absorption, k_B is the wavevector of the acoustic wave, ρ is the average density of the medium, and $\gamma = c_p/c_V$ is the ratio of principal specific heats at constant pressure c_p and at constant volume c_V.

In a polyatomic gas, further loss mechanisms can also be present due to the slow exchange of translational energy into vibrational and rotational modes, at high acoustic frequencies. A polyatomic gas carries energy in both external (translational) and internal (vibrational and rotational) degrees of freedom. In an acoustic wave, the periodic variation of temperature first affects the translational degrees of freedom, followed by a slower exchange of energy with the internal degrees of freedom via collisions between the gas molecules. As a result, the thermodynamic properties of the gas acquire frequency dependence. For example, the static value of the specific heat is composed of all its activated degrees of freedom:

$$c_V = c_{tr} + c_{vib} + c_{rot}. \qquad (2.5)$$

At high acoustic frequencies, the effective specific heat is complex and gives rise to dispersion and absorption of the acoustic wave. The absorption mechanism can be described as a relaxation process, and it has been shown experimentally that in the majority of cases the absorption can be characterized by single relaxation times for the vibrational modes (τ_v) and rotational modes (τ_r). The relaxation times represent the characteristic time required to equipartition energy by collisions into the internal modes.

For a heavy polyatomic gas molecule, such as SF_6, the major contribution to the specific heat arises from the vibrational modes and the relaxation times lie typically in the region $\tau_v \approx 10^{-7}$ s, $\tau_r \approx 10^{-10}$ s, at 1 atm. The relaxation times are proportional to the collision time and scale as the reciprocal of the gas density ($\tau_v \propto 1/\rho$).

The acoustic frequency (ν_B) generated in stimulated Brillouin scattering is given by

$$\nu_B = 2n \frac{v}{c} \nu_L \qquad (2.6)$$

where n is the refractive index of the medium, c is the speed of the light, v is the acoustic velocity of hypersound, and ν_L is the laser frequency. In the heavy gases, the acoustic frequency produced by SBS using visible radiation is ~500 MHz, and at high gas pressures this hypersonic frequency can lie close to a region of strong attenuation by vibrational relaxation (i.e. $\nu_B \approx 1/\tau_v$). The contribution to the acoustic linewidth from this term is given by [2.42]

$$\Gamma_{vib} = \frac{c_{vib}(c_p - c_v)}{\sqrt{c_p c_v (c_p - c_{vib})(c_v - c_{vib})}} \frac{\omega_B^2 \tau_v'}{1 + (\omega_B \tau_v')^2} \qquad (2.7)$$

where the vibrational component of the specific heat is given by

$$C_{vib}(T) = \sum_i g_i R \left(\frac{h\nu_i}{kT} \right)^2 \frac{\exp(h\nu_i/kT)}{(\exp(h\nu_i/kT) - 1)^2} \qquad (2.8)$$

and

$$\tau_v' = \sqrt{\frac{(c_p - c_{vib})(c_v - c_{vib})}{c_p c_v}} \tau_v \qquad (2.9)$$

where g_i and ν_i are the degeneracy and frequency of the ith vibrational state, $\omega_B = 2\pi\nu_B$ and R is the gas constant.

The rotational relaxation time is fast compared with the vibrational time and, typically, only requires a few binary collisions Z_r to equilibrate with the translational temperature (e.g. in SF_6, $Z_r \cong 4$). The rotational damping follows the same density dependence as the damping by viscosity. It is normally incorporated into the damping equation by introducing a bulk viscosity (η_B) that, in polyatomic gases, is given by

$$\eta_B = 0.131\eta Z_r. \qquad (2.10)$$

The overall equation for acoustic decay including classical and relaxation contributions can be written as

$$\Gamma_B = \tau_B^{-1} = \frac{k_B^2}{\rho}\left[\frac{4}{3}\eta + \eta_B + \frac{(\gamma-1)\kappa}{c_p}\right]$$

$$+ \frac{c_{vib}(c_p - c_v)}{\sqrt{c_p c_v (c_p - c_{vib})(c_v - c_{vib})}} \frac{\omega_B^2 \tau_v'}{1 + (\omega_B \tau_v')^2}. \qquad (2.11)$$

To a good approximation in many gases, one can also relate thermal conductivity to the viscosity by the Eucken relationship [2.15]

$$\frac{\kappa}{c_p} = \eta \frac{(9\gamma - 5)}{4\gamma}. \qquad (2.12)$$

Ideally, the experimentally determined values for κ at different pressures should be used but, since complete data of many heavy Brillouin-active gases are not available, equation (2.12) is assumed. Equation (2.11) can thereby be written in a simplified form for polyatomic gases using $\gamma = \frac{4}{3}$ and equation (2.12):

$$\tau_B^{-1} = 1.77 \frac{k_B^2}{\rho}\eta(1 + 0.074 Z_r)$$

$$+ \frac{c_{vib}(c_p - c_v)}{\sqrt{c_p c_v (c_p - c_{vib})(c_v - c_{vib})}} \frac{\omega_B^2 \tau_v'}{1 + (\omega_B \tau_v')^2}. \qquad (2.13)$$

For the case $\omega_B \tau_v' \gg 1$, the vibrational term varies as $1/\tau_v'$. The vibrational decay term is approximately proportional to density and will increase at higher pressure (unlike the viscosity term, which decreases as ρ^{-1}). Hence, at higher gas pressures, the vibrational contribution can become the major decay process. The acoustic decay time will therefore not always increase monotonically with pressure, but can rise to a maximum and decrease at higher pressures. These results indicate the possibility of achieving both high Brillouin-gain and fast response time (τ_B), at high pressures in suitable polyatomic gases.

2.3 Determination of SBS properties in gases

In many media the acoustic decay time is inferred by measurement of the Brillouin linewidth. The long acoustic decay time of high-pressure gases, however, allows a direct method of experimental determination of τ_B, as was performed in [2.42]. A laser system for measuring the acoustic decay time is shown in figure 2.1, consisting of a single-longitudinal and single-transverse TEM$_{00}$ mode ruby laser with pulse duration of 40 ns. The laser pulse was passed through an electro-optic shutter consisting of a Pockels

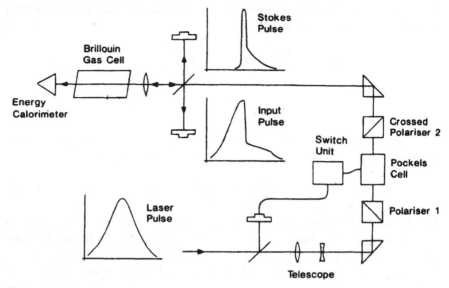

Figure 2.1. Experimental arrangement for measuring the acoustic decay time of SBS hypersound.

cell held at half-wave voltage placed between a pair of crossed polarizers. This initially transmits the incident laser pulse. A photodiode monitoring the incoming pulse was used to switch the Pockels cell voltage to ground potential, and rapidly blocking transmission of the back of the laser pulse. By partially uncrossing the second polarizer, a low level of light leakage was induced into the transmitted pulse after rapidly switching off the Pockels cell. The shaped pulse was focused by a lens ($f = 10\,\text{cm}$) into the SBS gas medium in a stainless steel cell ($l = 9\,\text{cm}$) whose windows were offset at a small angle to eliminate spurious reflections into the backscattering direction. By this relatively simple technique, the front of the laser pulse acts as an intense pump to generate the hypersonic wave by SBS in the gas cell, while the tail of the pulse behaves as a weak probe to monitor its subsequent decay.

It can be readily shown that the probe reflectivity is directly proportional to the acoustic intensity at low probe reflectivity. Consider the interaction of a plane-wave input laser field $E_L(z, t)$ (the analysis can be extended to the focused case used in the experiment). The acoustic field $\rho(z, t)$ is generated by the interaction of the laser field, E_L and backscattered Stokes field, E_S and can be described by

$$\frac{\mathrm{d}\rho}{\mathrm{d}t} = aE_L E_S^* - \rho/2\tau_B \tag{2.14}$$

where a is a coupling constant.

After the Pockels cell has switched off, the weak probe field, E_L must satisfy the condition

$$aE_L E_S^* \ll \rho/2\tau_B. \tag{2.15}$$

In this regime, the solution for the acoustic field at subsequent times is $\rho(z,t) = \rho(z,0)\exp(-t/2\tau_B)$. Also, if the amplitude of the probe input field does not vary significantly during the Stokes propagation time through the interaction cell (a good approximation when using a short cell of length l), then the output Stokes field $E_S(0,t)$ is given by

$$E_S(0,t) = b \int_l^0 \rho^*(z,t)E_L(z,t)\,dz. \tag{2.16}$$

Here, $g_B = 4ab\tau_B$ is the steady-state Brillouin-gain coefficient and $I_L = |E_L|^2$ is the laser intensity.

For the case with the weak probe, the field in the medium can be expressed as $E_L(z,t) = E_L(0,t)f(z,t)$, where $f(z,t)$ represents the transmission factor through the medium resulting from probe depletion in the SBS process. The reflectivity of the probe is therefore given by

$$R(t) = \left| \frac{E_S(0,t)}{E_L(0,t)} \right|^2 = \exp(-t/\tau_B)\left| b \int_l^0 \rho^*(z,0)f(z,t)\,dz \right|^2. \tag{2.17}$$

Hence, when the probe depletion is small, $f(z,t) = 1$ and

$$R(t) = R(0)\exp(-t/\tau_B) \tag{2.18}$$

the reflection coefficient of the probe intensity directly follows the decaying acoustic intensity and has a decay time constant equal to the acoustic decay time τ_B.

In the experiment, the input and backscattered signals were monitored on fast photodiodes, and from an analysis of the decaying reflectivity measurements, the acoustic decay time was directly deduced. To achieve a low reflectivity ($<1\%$) throughout the entire probing tail of the pulse, the pump energy was maintained close to threshold. An additional experimental aid for temporally synchronizing the two monitored signals was produced by inducing a small periodic modulation of the voltage across the Pockels cell after switching.

A calorimeter placed behind the gas cell monitored the laser energy and provided a method of measuring the relative Brillouin-gain coefficient g_B using the transient gain equation [2.42]. Using the measured transient threshold energy (U_{th}) and decay time (τ_B) yields a relative measure of the gain coefficient (g_B) from

$$\frac{g_B U_{th}}{\tau_B} = \text{constant} \tag{2.19}$$

when a constant interaction geometry for the experimental measurements is assumed.

The acoustic frequency, ν_B, can be measured by comparing the laser and Stokes frequencies using a Fabry–Pérot etalon. An alternative temporal method can also be performed by heterodyning the backscattered Stokes radiation with the input laser field and observing the beat frequency. In the investigated gases (Xe, SF_6, C_2F_6, $CClF_3$), the acoustic frequency, ν_B, is in the range 200–600 MHz and is resolvable on a fast oscilloscope. The evidence for gas breakdown could also be obtained in this experiment, by observing the degradation or premature termination of SBS, together with absorption of the laser radiation passing through the gas cell.

2.3.1   Xenon

Xenon is a monatomic gas that can be completely described by the classical damping formula as given in equation (2.4). Figure 2.2 shows the experimentally determined acoustic decay time as a function of xenon pressure using the optical probing technique for SBS excited at the ruby laser wavelength, $\lambda = 694.3$ nm. Errors on the decay time measurements were typically better than 6%. Experimental values of τ_B were extracted from measurements of reflectivity versus time by using a least-squares fit to the decaying reflectivity of the probing tail of the laser pulse. The solid line on the graph is plotted according to equation (2.4) using available thermodynamic data on xenon. At the pressures used in this heavy gas, the ideal gas behaviour is no longer valid, so virial coefficients for evaluating the gas density and the

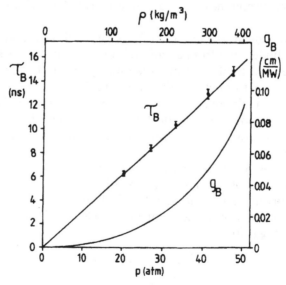

Figure 2.2. Acoustic decay time and gain coefficient in xenon (solid line shows calculated decay times assuming classical damping).

experimentally tabulated pressure dependence of the viscosity were used in the calculations for the real gas [2.41]. The agreement between the calculated and experimental decay time in xenon is very close.

From knowledge of the decay time τ_B, the Brillouin-gain coefficient g_B is calculated as

$$g_B = \frac{k_S^2 (\gamma_e)^2}{n c \nu_B \rho} \tau_B \qquad (2.20)$$

where k_S is the wave vector of the Stokes field, $\gamma_e = \rho(\partial p/\partial \rho)_T$ is the electrostrictive coefficient, n is the refractive index, and ν_B is the acoustic velocity of hypersound. The gain coefficient g_B is also plotted in figure 2.2. The relative variation of the gain with pressure, as estimated from the transient energy threshold, was consistent with the variation of the calculated gain. The results for this gas are also consistent with previously published values $g_B = 0.044$ cm/MW, $\tau_B = 65$ ns at $p = 50$ atm and 1.315 μm [2.41].

Focal intensities of several GW/cm^2 were used in the experiments without any major gas breakdown observed. This is in contrast to experiments at 1.315 μm in Xe [2.41], in which breakdown was observed at intensities of only 170 MW/cm^2, although the pulse duration was of the order of 10 μs in that case.

Using the calculated and experimental values of the decay time and its frequency dependence according to equation (2.4), an approximate expression for τ_B as a function of pressure p (atm) and laser wavelength λ_L (μm) was found [2.42]:

$$\tau_B \text{ (ns)} = 0.65 \lambda_L^2 p. \qquad (2.21)$$

2.3.2 Sulphur hexafluoride (SF$_6$)

Sulphur hexafluoride is a heavy polyatomic gas that has been used frequently as a high-gain Brillouin-active medium operating at modest pressures [2.41] compared with other lighter gases. Unlike xenon, it has a large number of vibrational and rotational degrees of freedom. To calculate the decay time in this gas, it is necessary to use the complete equation (2.11), that includes vibrational and rotational damping. Fairly complete data of the thermodynamic properties and also relaxation times from ultrasound studies exist for SF$_6$. The density, viscosity and velocity dependence with pressure were also incorporated using available data [2.42].

Figure 2.3 shows the experimentally determined decay time in SF$_6$ together with its calculated variation with pressure up to its saturated vapour pressure. The decay time according to the absorption equation (2.4) is also included (dashed line) for comparison. The experimental and theoretical determination of τ_B are again in good agreement and the significant reduction from the classical value at higher pressures should be noted.

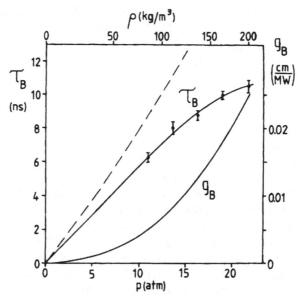

Figure 2.3. Acoustic decay time and gain coefficient in SF$_6$ (solid line shows calculated values; dashed line shows classical decay values).

An approximate expression has been derived to estimate the acoustic decay time in SF$_6$ as a function of density at other wavelengths using the values of the calculated decay time [2.42], that is,

$$\tau_{\rm B}^{-1} = \frac{5.9 \times 10^9}{\rho \lambda_{\rm L}^2} + 1.6 \times 10^5 \rho \qquad (2.22)$$

where ρ is in kg/m^3 and $\lambda_{\rm L}$ is in μm.

Using the experimental values of $\tau_{\rm B}$ in equation (2.20), the gain coefficient for SF$_6$ has also been calculated and is shown in figure 2.3. It is noted that, at higher pressures, its value is slightly less than previously quoted values in the literature [2.41] ($g_{\rm B} = 0.035$ cm/MW at 22 atm).

Gas breakdown was observed at an estimated focal intensity of ~1 GW/cm^2, in qualitative agreement with values quoted at 1 μm (5 GW/cm^2) and 249 nm (10 GW/cm^2) [2.43].

Although SF$_6$ is a heavy gas with high Brillouin-gain, its range of operation is limited by its saturated vapour pressure, which is approximately 22 atm at room temperature. An increase in gas pressure and hence gain can only be achieved by increasing the temperature. In a survey of the properties of several heavy gases, in order to select those with possible high Brillouin gain and the capability of high-pressure operation at room temperature [2.42], two promising candidates, chlorotrifluoromethane (CClF$_3$) and hexafluoroethane (C$_2$F$_6$), were selected for experimental characterization. Both $CClF_3$ and C_2F_6 are gases with high density, and hence correspondingly high Brillouin gain. They also exhibit low-lying vibrational levels, an indication that the gases will have a fast vibrational relaxation time and hence strong vibrational damping at higher pressures.

2.3.3 Chlorotrifluoromethane ($CClF_3$)

$CClF_3$ is a heavy polyatomic gas with a saturated vapour pressure of approximately 32 atm ($T = 20\,°C$). The acoustic decay time was measured experimentally and calculated using available data of the thermodynamic and relaxation properties of the gas. The characteristic number of collisions for energy transfer to rotational modes was estimated by comparison with similar molecules as $Z_r = 3$, although the value of the decay time is insensitive to the precise choice of this number. The Brillouin-gain coefficient was also derived for this gas according to equation (2.21). The results are displayed in figure 2.4. Strong vibrational relaxation is observed in this gas at higher pressures such that the decay time rises to a maximum 9 ns at a pressure of approximately 27 atm and decreases at higher pressures. The high Brillouin-gain coefficient is also confirmed by the energy threshold measurements.

Gas breakdown in $CClF_3$ was not observed even at the highest pumping intensity from the ruby laser system ($\sim10\,GW/cm^2$). A reflection coefficient

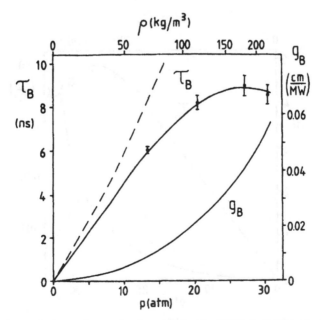

Figure 2.4. Acoustic decay time and gain coefficient in $CClF_3$ (solid line shows calculated values; dashed line shows classical decay values).

of greater than 70% was obtained under these conditions. (A reflectivity of ~95% was also achieved in this gas using a high-power Nd:YAG laser system at 1.06 μm and focal intensities ~100 GW/cm² without observation of breakdown.)

An approximate expression has been derived for the acoustic decay time as a function of gas density in kg/m³ and laser wavelength in μm [2.42]:

$$\tau_{\mathrm{B}}^{-1} = \frac{4.78 \times 10^9}{\rho \lambda_{\mathrm{L}}^2} + 3.25 \times 10^5 \rho. \qquad (2.23)$$

2.3.4   Hexafluoroethane (C_2F_6)

Unlike SF_6, $CClF_3$ and the majority of the heavy fluorinated compounds, hexafluoroethane (C_2F_6) at normal room temperature lies above its critical temperature ($T_c = 19.7\,^\circ C$ [2.29]). Since the gas cannot liquefy above its critical temperature, it can be used at high pressure. Its thermodynamic properties were not as readily available compared with the other gases investigated, and no vibrational or rotational relaxation times have been measured. The relaxation times can, however, be estimated by comparison with other fluorinated molecules.

An experimental determination of the decay time of C_2F_6 was made [2.42] and is plotted in figure 2.5. A pronounced variation in the decay

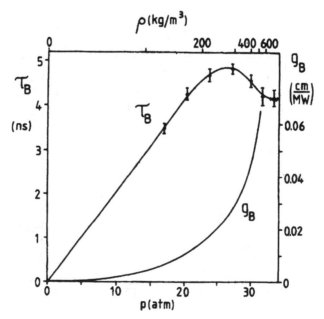

Figure 2.5. Experimentally measured acoustic decay times and derived gain coefficients in C_2F_6.

time is observed in the vicinity of its critical density and is caused by the very large variation in density with pressure $(dp/d\rho)_T \to 0$ in this region. The unusual behaviour of the gas in its critical region will be illustrated again in the section on the measurement of the acoustic velocity.

2.3.5 Liquid CClF$_3$

Most of the fluorocarbon compounds have the useful characteristic of high density, low viscosity, chemical stability, and transparency from the infrared to ultraviolet regions of the spectrum. These properties are present both in their gaseous and liquid phases.

The gas CClF$_3$ was liquefied, at a pressure near its saturated vapour, by adding quantities of high-pressure xenon. The acoustic decay time of the liquid was measured as 6.2 ± 0.4 ns, at 32 atm. The decay time is relatively long compared with other liquids, primarily due to the low viscosity of the liquid CClF$_3$. Indeed, the decay time is comparable with that of CClF$_3$ at high pressure in its gaseous phase.

According to equation (2.6), the Brillouin frequency shift is determined by the acoustic velocity (v_B) and the refractive index (n) of the medium. In a polyatomic gas, at typical hypersonic frequencies (500 MHz), the adiabatic velocity $v_B^2 = \gamma_\infty RT/M$ is determined by the high-frequency value of the ratio of specific heats $\gamma_\infty = (C_p - C_{vib})/(C_v - C_{vib})$, in which the vibrational contribution to the specific heats is not activated, and by the molecular mass, M. The hypersonic velocity is therefore higher than the low-frequency acoustic velocity. In a real gas, the acoustic velocity is also a function of pressure. For example, in SF$_6$, the dependence of the hypersonic velocity at $T = 26\,°C$ with density is measured to be [2.42]

$$v_B = 149.9 - 0.9876\rho - 6.386 \times 10^{-4}\rho^2 \pm 1.3 \quad \text{m/s}. \qquad (2.24)$$

The acoustic velocity can be experimentally measured for the gases such as CClF$_3$ and C$_2$F$_6$ by heterodyning the backscattered and input radiation, which gives a beat frequency equal to their difference frequency. The acoustic velocity was determined using the resultant beat frequency (v_B) in equation (2.6) and the results are depicted in figures 2.6 and 2.7. The acoustic velocity in a real gas is given by $\sqrt{(dp/d\rho)_{ad}}$, under adiabatic conditions. For CClF$_3$, as in all non-ideal gases, this is dependent on pressure, as seen in figure 2.6. For C$_2$F$_6$, the velocity variation versus pressure is rather more complex, which results from the fact that the operating temperature $(T = 21\,°C)$ was close to the critical point $(T_c = 19.7\,°C)$. At the critical point, $(dp/d\rho)_T = (d^2p/d\rho^2)_T = 0$ and one would predict a minimum in the acoustic velocity in this region. In figure 2.7, at $T = 21\,°C$, the minimum acoustic velocity occurs very close to the critical density and falls to a relatively low value. It is also noted that in the experiment, the C$_2$F$_6$ gas became visibly turbulent

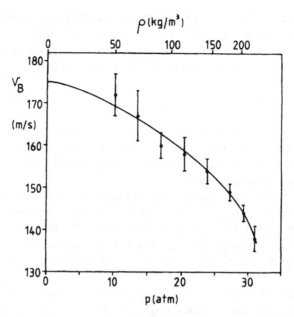

Figure 2.6. Experimentally measured acoustic velocities of hypersound in $CClF_3$.

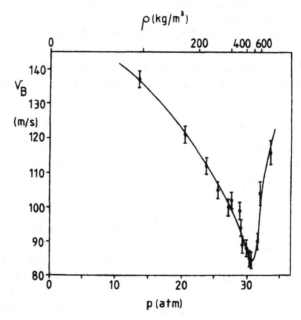

Figure 2.7. Experimentally measured acoustic velocities of hypersound in C_2F_6 in the vicinity of the critical point. Operating temperature $T = 21\,°C$ ($T_c = 19.7\,°C$, $P_c = 29.8\,atm$).

near its critical density due to the very large density fluctuations near the critical point (critical opalescence).

In these heavy gases, it is observed that the acoustic velocity and frequency shift are small compared with lighter gases or liquids (e.g. CH_4, $\nu_B = 440 \, m/s$; acetone $\nu_B = 1190 \, m/s$). The small acoustic frequency is potentially useful for several applications such as in aberration correction in laser amplifiers with very narrow gain linewidth, e.g. the iodine laser [2.41].

2.4   Determination of SBS properties of liquid materials

The liquid SBS materials can be investigated with similar methods as gaseous materials, although shorter acoustic decay times and higher acoustic frequencies make direct temporal measurements harder to perform. For best performance, the materials should be ultra-pure, usually vacuum distilled several times to remove the particulates in the liquids. The last distillation could be done directly in the SBS cell, when the cleanliness of the ensemble is observed by the scattering of a visible c.w. laser beam passed through the cell. The absence of particles in liquid materials proved critical to thermal lensing, breakdown or self-focusing at intensities higher than a thousand times their SBS threshold [2.8, 2.12, 2.13].

A typical simplified experimental setup, in which the stimulated Brillouin scattered light is measured, is shown in figure 2.8 [2.12, 2.13, 2.46].

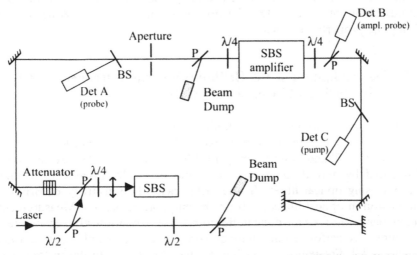

Figure 2.8. Experimental setup for parameter measurement of SBS liquids [2.12, 2.13, 2.46]. SBS: focused cell; SBS: amplifier cell (both filled with common SBS liquids); P: polarizers; Det A, Det B, Det C: calibrated energy detectors; $\lambda/4$: quarter-wave plates; $\lambda/2$: half-wave plates; BS: beam splitters.

The laser beam is split into a pump beam and a probe beam, for the gain measurement. The probe beam is backscattered by the SBS cell #1 (in focused geometry) and then arrives simultaneously with the pump beam to the SBS cell #2 (amplifier). Calibrated calorimeters CPU, CP and CS measure the energy of the pump, probe and scattered pulses, respectively. The SBS cell is temperature controlled, which allows thermal tuning of narrowband Brillouin lines. The Brillouin frequency varies proportionally with temperature. The Brillouin-gain coefficient, g_B, is calculated from the probe energy at the entrance and at the exit of the SBS amplifier cell (#2), E_{in} and E_{out} respectively, by

$$g_B = \frac{\ln(E_{out}/E_{in})}{LI_p(0)} = \frac{At_p \ln(E_{out}/E_{in})}{2LE_p} \tag{2.25}$$

where $I_p = 2E_p/At_p$ is the pump intensity at the beam centre, E_p is the total pump energy, A is the pump beam area (at e^{-2} intensity point cross-section, which could be measured from an image captured by a CCD camera), t_p is the respective pulse duration and L is the length of SBS amplifier cell. Scanning the temperature of SBS cell #1, from about the room temperature, by some tens of degrees and keeping the temperature of SBS amplifier cell (#2) constant, the probe frequency was scanned by some hundreds of MHz. This range is enough to measure $E_{out}(\nu_B)$ and to calculate $g_B(\nu_B)$ with equation (2.25). From a best fit of these data with a Lorentzian function, one can deduce the Brillouin-gain coefficient at the line centre, g_B and the full gain linewidth at half-maximum, $\Delta\nu_B = \pi\tau_B$. Using these measurements, Erokhin et al [2.13] characterized precisely SBS properties of various SBS liquids, Amimoto et al [2.12], the properties of $SnCl_4$, and Watkins et al [2.46], the properties of $TiCl_4$.

2.5   Determination of the Brillouin linewidth, frequency and gain coefficient in solid materials

High resolution stimulated Brillouin-gain spectroscopy in glasses and crystals was done in many studies, particularly by Pohl and Kaiser [2.14], Faris et al [2.16] and more recently Le Floch et al [2.47]. This research was stimulated by optical communications and by Brillouin-based distributed sensors using optical fibres. In the latter case, the sensitivity of Brillouin interaction in optical fibres to strain or temperature is used. Measurements at low temperatures were done in liquid nitrogen and those above room temperature were performed in a temperature-controlled chamber (up to about 350 K) [2.47]. For temperatures higher than 150 K, the Brillouin frequency increases linearly with temperature in standard single-mode optical fibres (with the slope $1.31\,MHz\,K^{-1}$, at $\lambda_0 = 1.55\,\mu m$), which corresponds to the observations of many authors. This linear dependence allows a simple

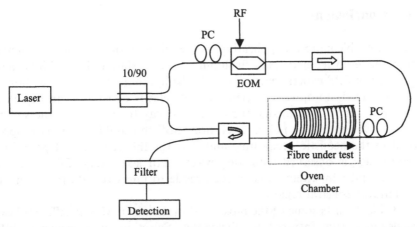

Figure 2.9. Experimental setup for SBS parameter measurement in an optical fibre [2.47]. PC: polarization controller; EOM: electro-optical modulator biased at its zero-transmission point; Filter: blocking the Rayleigh scattering and the anti-Stokes signal.

linear scanning of probe signal frequency and the calculation of the Brillouin-gain spectrum.

A simple experimental setup (Brillouin optical time-domain analyser) for the study of the Brillouin-gain spectrum is shown in figure 2.9 [2.47]. The light source is a semiconductor laser with a power of 30 mW, at $\lambda_0 = 1.55\,\mu$m. The laser beam passes through a polarization controller (PC) and an electro-optical modulator (EOM) biased at its zero-transmission point. This modulation leads to two side-bands, which could act as a probe, around the laser line which acts as the pump beam (I_p). The probe is backscattered through the interaction, in the fibre, with the counter-propagating pump wave (at a power level of about 20 mW). The anti-Stokes wave is depleted and the Stokes wave (I_S) is amplified in this process, which could be described, in the steady-state, by the well-known coupled equations

$$(dI_p/dz) = -\alpha I_p - g_B I_S I_p, \qquad (dI_S/dz) = -\alpha I_S - g_B I_S I_p \qquad (2.26)$$

where α is the absorption coefficient, $g_B = g_{B0}[1 + 2(\nu - \nu_B)^2/(\Delta\nu_B)^2]^{-1}$ is the frequency-dependent Brillouin-gain coefficient, ν_B is the resonant Brillouin frequency shift and $\Delta\nu_B$ is the Brillouin linewidth. Assuming a negligible depletion of the pump beam, one can find the classical solution of the exponential growing Stokes signal, which could indeed be detected experimentally, after the filter cutting the Rayleigh scattering and the anti-Stokes signal. From the Lorentzian shape and centring of the Brillouin-gain spectrum, one can determine ν_B, $\Delta\nu_B$, and g_{B0}. For a standard single-mode fibre, $\nu_B = 10.842\,$MHz and $\Delta\nu_B \approx 60\,$MHz.

2.6 Conclusions

The first problem in SBS applications is the selection of a nonlinear material and of the setup geometry. The Brillouin frequency shift and linewidth can be measured in SBS experiments. Then, the Brillouin-gain factor, the phonon lifetime, the elasto-optic coefficient and the elastic constants can be calculated. For simple SBS media, the acoustic decay time scales as the square of the wavelength. For pulsed operation, SBS threshold energy is proportional to the acoustic decay time and it is for this reason that poor SBS performance is observed with long wavelength lasers (e.g. CO_2 lasers at 10.6 μm) and why attempts have been made to improve SBS media in the mid-infrared spectral region.

Table 2.1 lists some of the most used materials in SBS at different laser wavelengths from infrared to ultraviolet. Selection of appropriate SBS material depends on several factors some of which are characterized by the displayed SBS material parameters and by the cited papers, which give more references for the interested reader.

In general, the use of pressurized gases eliminates complications arising from liquid use including corrosion, cavitation bubbles and thermal convective turbulence. The use of liquids is often selected, however, due to lower SBS thresholds and larger Brillouin linewidths (one order of magnitude shorter acoustic decay times, i.e. sub-nanoseconds). The solid SBS materials are used for all-solid state applications, mostly in guiding geometry (optical fibres), in which their small Brillouin-gains are offset by long interaction lengths.

In order to select the optimum material for a particular experimental system, it is important to have a reliable characterization of SBS materials. The criteria for choosing a suitable material depend on laser wavelength and pulse duration. Some straightforward optical methods have been described whereby the main parameters of SBS can be directly measured. Generally, good agreement is found between the experimental and theoretically calculated values of the Brillouin frequency shift, linewidth and gain coefficient, in gases, liquids and solids.

References

2.1 Kaiser W and Maier M 1972 Stimulated Rayleigh Brillouin and Raman spectroscopy in *Laser Handbook* ed F T Arecchi and E O Schultz-Dubois (Amsterdam: North-Holland) p 1078

2.2 Reintjes J F 1995 Stimulated Raman and Brillouin scattering in *Handbook of Laser Science and Technology* Supplement 2: *Optical materials* ed M J Weber (Boca Raton, FL: CRC Press) p 334

2.3 Ragul'ski V V 1990 *Phase Conjugation by SBS* (Moskow: Nauka Press)

2.4 Rockwell D A 1994 Application of phase conjugation to high-power lasers in *Optical Phase Conjugation* ed M Gower and D Proch (Berlin: Springer); 1988 *IEEE J. Quantum Electron.* **24** 1124

2.5 Duignan M T, Feldman B J and Whitney W T 1987 *Opt. Lett.* **12** 111

2.6 Whitney W T, Duignan M T and Feldman B J 1990 *J. Opt. Soc. Am. B* **7** 2160

2.7 Vorontsov E N *et al* 1990 *Sov. J. Quantum Electron.* **20** 256

2.8 Eichler H J, Haase A, Liu B and Mehl O 1998 Phase conjugation techniques in *Optical Resonators—Science and Engineering* ed R Kossowsky *et al* (Dordrecht: Kluwer) p 103

2.9 Zykov L I, Buyko S A, Dolgopolov Yu V, Dudov A M, Eroshenko V A, Kirillov G A, Kochemasov G G, Kulikov S M, Novikov V N, Sharpa A F and Sukharev S A 1998 SBS properties of high-pressure xenon at density of 0.3–1 g/cm3 in *Optical Resonators—Science and Engineering* ed R Kossowsky *et al* (Dordrecht: Kluwer) p 479

2.10 Dolgopolov Yu V, Kulikov S M and Solov'yeva M M 1988 *Bull. Acad. Sci. USSR Phys.* **52** 549

2.11 Bessarab A V *et al* 1988 *Bull. Acad. Sci. USSR Phys.* **52** 333

2.12 Gross R Q F, Amimoto S T and Garman-DuVall L 1991 *Opt. Lett.* **16** 94; 1991 *Opt. Lett.* **16** 1382

2.13 Erohkin A I, Kovalev V I and Faizullov F S 1986 *Sov. J. Quant. Electron.* **16** 872; 1990 *Sov. J. Quant. Electron.* **16** 1267

2.14 Pohl D and Kaiser W 1970 *Phys. Rev. B* **1** 31

2.15 Denariez M and Bret G 1968 *Phys. Rev.* **171** 160

2.16 Faris G W, Jusinski L E and Hickman A P 1990 *J. Opt. Soc. Am. B* **10** 587

2.17 Narum P, Skeldon M D and Boyd R W 1986 *IEEE J. Quant. Electron.* **QE-22** 2161

2.18 Hwa L G, Schroeder J and Zhao X S 1989 *J. Opt. Soc. Am. B* **6** 833

2.19 Slatkine M, Bigio I J, Feldman B J and Fisher R A 1982 *Opt. Lett.* **7** 108

2.20 Armandillo E and Proch D 1983 *Opt. Lett.* **8** 523

2.21 Pfau A K, Proch D and Bachmann F 1990 *Opt. Lett.* **15** 6

2.22 Osborne M R, Schroeder W A, Damzen M S and Hutchinson M H R 1989 *Appl. Phys. B* **48** 351

2.23 Nassisi V and Pecoraro A 1993 *IEEE J. Quantum Electron.* **QE-29** 2547

2.24 Filippo A A and Perrone M R 1992 *IEEE J. Quantum Electron.* **QE-28**; 1993 *Appl. Phys. B* **57** 103

2.25 Alimpiev S S *et al* 1990 *Sov. J. Quantum Electron.* **20** 276

2.26 Bigio I J, Feldman B J, Fisher R A and Slatkine M 1981 *IEEE J. Quantum Electron.* **QE-17** 220

2.27 Gower M C and Caro R G 1982 *Opt. Lett.* **7** 162

2.28 Gower M C 1983 *Opt. Lett.* **8** 70

2.29 Andreev N, Kulagin O, Palashov O, Pasmanik G and Rodchenkov V 1995 *Proc. SPIE* **2633** 476

2.30 Krainov V V, Mak A A, Rusov V A and Yashin V E 1991 *Kvant. Elektron.* **18** 959

2.31 Moore G T 2001 *IEEE J. Quantum Electron.* **QE-37** 781

2.32 Sutherland R L 1996 *Handbook of Nonlinear Optics* (New York: Marcel Dekker) p 629

2.33 Buiko S A, Kulikov S M, Novikov V N and Sukharev S A 1999 *Zh. Eksp. Teor. Fiz.* **89** 1051

2.24 Bespalov V G, Efimov Y N and Staselko D I 1998 *Optics and Spectroscopy* **85** 878

2.35 Kim H S, Ko D K, Lim G, Cha B H and Lee J 1999 *Opt. Commun.* **171** 177

2.36 Hsu H and Li T N 2000 *Appl. Opt.* **39** 6528

2.37 Lee C C and Chi S 2000 *IEEE Photonics Technol. Lett.* **12** 672

2.38 Yoshida H, Fujita H, Nakatsuka M and Yoshida K 1999 *Jpn. J. Appl. Phys.* **38** L521

2.39 Kovalev V I, Popovichev V I, Ragul'skii V V and Faizullov F S 1972 *Sov. J. Quantum Electron.* **2** 69 (1972 *Kvant. Elektron.* **9** 78)

2.40 Maier M and Renner G 1971 *Phys. Lett.* **34A** 299

2.41 Dolgopolov Yu V, Komarevskii V A, Kormer S B, Kochemasov G G, Kulikov S M, Murugov V M, Nikolaev V D and Sukharev S A 1979 *Sov. Phys. JETP* **49** 458 (1979 *Zh. Eksp. Teor. Fiz.* **76** 908)

2.42 Damzen M J, Hutchinson M H R and Schroeder W A 1987 *IEEE J. Quant. Electron.* **QE-23** 328

2.43 Goldhar J and Murray J R 1979 in *Proc. Conf. Laser Eng. Appl.* (IEEE-OSA) p 5

2.44 Zel'dovich B Ya, Pilipetskii N F and Shkunov V V 1982 *Sov. Phys. Usp.* **25** 713 (1982 *Usp. Phys. Nauk* **138** 249)

2.45 Basov N G, Efimkov V F, Zubarev I G, Kotov A V, Mironov A B, Mikhailov S I and Smirnov M G 1979 *Sov. J. Quantum Electron.* **9** 455 (1979 *Kvant. Elektron.* **6** 765)

2.46 Watkins D E, Scott A M and Ridley K D 1990 *Opt. Lett.* **15** 1267

2.47 Le Floch S, Riou F and Cambon P 2001 *J. Opt. A: Pure Appl. Opt.* **3** L12

Chapter 3

Solutions of the one-dimensional SBS model

The simplest interaction geometry to consider is when the SBS process involves an incident pump beam with a plane-wavefront and the scattered Stokes wave is also considered to have a plane-wavefront. In this regime, the SBS process can be considered as a one-dimensional spatial interaction, transverse derivatives can be neglected and considerable mathematical simplification of the process is achieved. Many of the important physical results, such as reflectivity and pulse durations, can be approximated in this regime. With further simplifying approximations, fairly simple but useful relationships can be deduced analytically, and these results can be used to predict first-order approximations to experiments and in applications.

3.1 The steady state regime

3.1.1 Laser pump depletion included, absorption neglected

In this regime, expounded by Kaiser and Maier [3.1], we consider the backscattering regime of figure 3.1 with all time derivatives set equal to zero with the interaction time long compared with the phonon lifetime. We also consider neglect of medium absorption and use the steady state intensity equations:

$$\frac{dI_L}{dz} = -g_B I_L I_S, \qquad \frac{dI_S}{dz} = -g_B I_L I_S. \qquad (3.1)$$

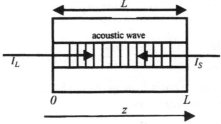

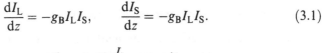

Figure 3.1. Backscattering geometry for SBS.

The analytical solution of Stokes intensity generated ($I_S(0)$) as a function of incident pump intensity $I_L(0)$ and interaction length L, with inclusion of pump depletion but neglect of medium absorption is given by

$$\frac{I_S(L)}{I_S(0)} = \frac{1 - I_S(0)/I_L(0)}{\exp\{[1 - I_S(0)/I_L(0)]g_B I_L(0)L\} - I_S(0)/I_L(0)} \tag{3.2}$$

with the additional conservation condition

$$I_S(z) - I_L(z) = I_S(0) - I_L(0) = \text{const.} \tag{3.3}$$

We can also specify an additional boundary condition from the empirical relationship relating to the spontaneous scattering that acts as an initiating source $I_S(L, t) = I_L(0, t)\,e^{-G_0}$, where $G_0 \approx 30$ is assumed typically. With this additional boundary condition, equation (3.2) can be rearranged to give a transcendental equation for the SBS reflectivity $R = I_S(0)/I_L(0)$:

$$G = g_B I_L(0)L = \frac{30 + \ln R}{1 - R}. \tag{3.4}$$

This result demonstrates that the SBS reflectivity is determined by the dimensionless exponential parameter $G = g_B I_L(0)L$. This expression allows calculation of the intensity requirement for a given reflectivity. For example, for $R = 1\%$ requires $G = 25.7$, for $R = 10\%$ requires $G = 30.8$ and for $R = 90\%$ requires $G = 299$. In comparison, the Stokes growth solution with neglect of pump depletion predicts

$$I_S(0) = I_S(L)\exp[g_B I_L(0)L] \tag{3.5}$$

and hence $R = \exp(G - 30)$ or $G = 30 + \ln R$. For $G = G_0 = 30$, the predicted reflectivity is 100% and is clearly wrong; for high reflectivities equation (3.4) must be used.

3.1.2   Laser intensity undepleted, absorption included

The laser intensity experiences exponential attenuation due to the absorption of the medium while the Stokes intensity sees SBS gain and absorption losses

$$\frac{dI_L}{dz} = -\alpha I_L, \qquad \frac{dI_S}{dz} = -g_B I_L I_S + \alpha I_L. \tag{3.6}$$

The solution of the above equations gives the Stokes intensity, as a function of distance within the medium, as

$$I_S(z) = I_S(L)\exp[g_B I_L(0)(e^{-\alpha z} - e^{-\alpha L})/\alpha - \alpha(L - z)] \tag{3.6a}$$

where L is the length of the medium and $I_S(L)$ and $I_L(0)$ are the scattered (Stokes) and the laser intensity at opposite ends of the medium, respectively.

For $\alpha L \ll 1$ (optically thin medium), one obtains the exponential amplification typical for all stimulated scattering processes:

$$I_S(z) = I_S(L) \exp\{[g_B I_L(0) - \alpha](L - z)\}. \tag{3.7}$$

One obtains net amplification of the scattered light for $g_B I_L(0) > \alpha$ only.

3.2 Transient plane-wave solutions

Transient phenomena of SBS were studied theoretically in non-absorbing media and assuming no pump depletion. Kroll [3.2], Hon [3.3] and other authors have given a general solution of the coupled equations, where the partial time derivatives of first order are considered and the pump pulse or interaction time is short or comparable with the phonon life time (τ_B). For large gain, the growth of the Stokes wave intensity (taken as a step-function) is given by

$$I_S(z,t) = I_S(L,t) \exp\{2[g_B I_L(0)(L - z)t/\tau_B]^{1/2} - t/\tau_B\}. \tag{3.8}$$

According to equation (3.8), the amplification depends on the ratio t/τ_B. This dependence leads to a greatly reduced gain when the pulses are short compared with the phonon lifetime. The main growth factor is seen as a function of both space and time coordinates. The growth increases with the accumulated fluence of the laser input ($U_L(t) = I_L(0)t$) rather than the instantaneous intensity. Indeed, for generally shaped pulses, the term $I_L(0)t$ in equation (3.8) can be replaced by $U_L(t) = \int I_L(t')\,dt'$. After a sufficiently long time, the Stokes gain reaches the steady-state described by equation (3.5). This time to steady-state is given by

$$t \geq g_B I_L(0) L \tau_B. \tag{3.9}$$

For high SBS growths the transient regime persists for a time an order of magnitude greater than the decay time τ_B.

A transient threshold can be defined assuming the total intensity gain factor $I_S(0,t)/I_S(L,t) \approx \exp(30)$ is reached by the end of the pump pulse with duration t (and assuming the transit time of the medium is short compared with the interaction time). This condition gives (t_L is the duration of the pumping pulse)

$$2\left[\frac{g_B}{\tau_B} L \int_0^{t_L} I_L(t')\,dt'\right]^{1/2} - \frac{t_L}{\tau_B} \approx 30. \tag{3.10}$$

For a square pulse this gives a threshold gain G_{th} given by

$$G_{th} = (g_B I_L L)_{th} = \frac{1}{4} \frac{\tau_B}{t_L}\left(30 + \frac{t_L}{\tau_B}\right)^2. \tag{3.11}$$

The threshold condition of equation (3.11) indicates that short pulse durations require higher threshold intensity (assuming $t_L/t_B < 30$). At very short pulse durations, the pulse duration becomes comparable with the acoustic frequency and the SBS process has insufficient bandwidth to couple the process efficiently, with the consequence the SBS threshold rises even more sharply than expected by equation (3.11). In this case it is necessary to retain the second time derivative in the acoustic wave equation.

3.3   Numerical solutions

A further approach, to remove the restrictions of approximations made for analytical simplification, is to perform numerical solutions. The system of transient equations describes a more generalized system of interaction for laser (E_L) and Stokes (E_S) fields and the acoustic wave amplitude Q (in suitably scaled units):

$$\left(\frac{\partial}{\partial t} + \frac{c}{n}\frac{\partial}{\partial z}\right)E_L = -E_S Q, \qquad \left(\frac{\partial}{\partial t} - \frac{c}{n}\frac{\partial}{\partial z}\right)E_S = E_L Q^*$$

$$\left(\frac{\partial}{\partial t} + \frac{1}{2\tau_B}\right)Q = \gamma E_L E_S^*. \qquad (3.12)$$

This system of equations can be numerically solved [3.4] and it has been shown that such a system is suitable for solving the regime of SBS in which strong compression of duration of laser pulses can be achieved [3.3–3.7]. The effects of medium absorption can be readily incorporated in this system of equations. Unfortunately, the numerical solution must be performed by computer for each specific interaction condition and set of initial input parameters and does not give general results, although it is a useful way to illustrate specific examples.

3.4   Solving one-dimensional SBS using the characteristic equations

In the remainder of this chapter, we use the method of characteristic equations [3.8] to solve in the general analytical way the set of coupled differential equations describing the one-dimensional SBS process, which are derived from the Maxwell and Navier–Stokes equations including optical absorption. The expressions of pump- and Stokes-wave intensities, which are obtained accounting for the first-order time and spatial derivatives and for the arbitrary time dependence of the pump pulse, are more accurate in comparison with the previous approximated results (e.g. from [3.1], [3.2]). The results are derived for pump pulses shorter and longer than the phonon lifetime.

The Stokes pulse has a different duration to the laser pulse and the compression ratio in both these cases is also calculated. We redefine SBS steady-state regime in the saturation region of these solutions as a function of the pump wave intensity. This definition can provide also a sufficient condition for a stationary process. The theoretical results are checked with experimental data to show the validity of this general model.

The SBS process is modelled by the set of differential equations, neglecting the second-order derivatives with respect to the first-order ones.

The phase-matching relation holds (from conservation laws):

$$\varepsilon = \varphi_L - \varphi_S - \varphi_B = 0 \tag{3.13}$$

where $\varphi_L = \omega_L t + k_L z$, $\varphi_S = \omega_s t - k_s z$ and $\varphi_B = \Omega_B t + k_B z$ are phases associated with the optical pump, the optical scattered and the acoustical fields. Condition (3.13) takes place when the Navier–Stokes equation is purely deterministic; this means that the statistical instabilities are neglected.

We consider the system of equations (1.30a–c) which describes the SBS process:

$$\frac{\partial E_L}{\partial z} + \frac{n}{c}\frac{\partial E_L}{\partial t} + \frac{1}{2}\alpha E_L = \frac{i\omega_L}{4cn}\frac{\gamma_e}{\rho_0} E_S \rho$$

$$-\frac{\partial E_S}{\partial z} + \frac{n}{c}\frac{\partial E_S}{\partial t} + \frac{1}{2}\alpha E_S = \frac{i\omega_S}{4cn}\frac{\gamma_e}{\rho_0} E_L \rho^* \tag{3.14}$$

$$\frac{\partial \rho}{\partial t} + \frac{\Gamma_B}{2}\rho = \frac{i\gamma_e \varepsilon_0 K_B}{4v} E_L E_S^*.$$

We introduce the following conditions:

$$\frac{\gamma_e}{2cn}(I_L I_S)^{1/2} \gg \rho v^2, \qquad \frac{\partial \rho}{\partial t} \ll \frac{\Gamma_B}{2}\rho \tag{3.15}$$

where ρ is the density variation of the nonlinear medium, v is the velocity of the acoustical waves, γ_e is the electrostrictive coefficient, n the refractive index, c the speed of light and I_L, I_S the intensities of the pump and scattered fields. The first condition in equation (3.15) means that the optical gain is much higher than the acoustical one (which is valid for restricted values of pumping intensities)[*]

$$I_0 \gg \left(\frac{n\omega_B}{g_B c}\right)\left(\frac{\tau_B}{t_L}\right).$$

The second condition is neglecting the acoustic dynamics through an adiabatic approximation (the acoustical field can be considered approximately quasi-stationary).

---

[*]The first condition in equation (3.15) can be written in the equivalent form: $g_B z_c I_0 \gg \omega\tau$. Using the expression (3.19) for the characteristic length z_c, we obtain the conditions for the maximum pumping intensity: $I_0 \gg (n\omega_B/g_B c)(\tau_B/t_L)$ (for $t_L < \tau$) and $I_0 \gg n\omega_B/g_B c$ (for $t_L > \tau$). The first one covers both cases and limits the validity of the SBS equation system (3.16) to pumping intensities higher than a certain value $I_1 = (n\omega_B/g_B c)(\tau_B/t_L)$. For CS_2 as the nonlinear SBS medium, at $\lambda_L = 1.06\,\mu m$, $t_L = 3\,ns$, $\tau = 6\,ns$, this limit is: $I_1 = 16.6\,MW/cm^2$.

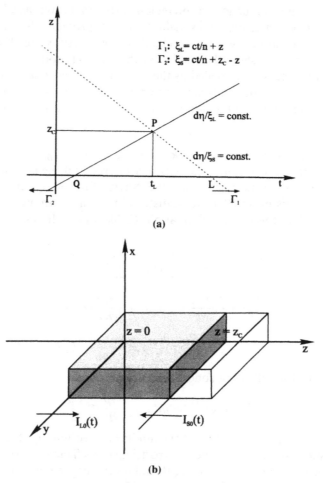

Figure 3.2. (a) The characteristic lines Γ_1 and Γ_2 associated with z_c and t_L are defined by equation (3.17); Q is the intersection of the Stokes characteristic with the time axis at $z = 0$; L is the intersection of the pump characteristic with the time axis at $z = 0$; P is the intersection of the Stokes and pump characteristics. (b) The dashed volume represents the intersection region between the optical and acoustic fields. I_{L0} is the intensity of the pump pulse defined at $z = 0$; I_{S0} is the intensity of the Stokes field defined at $z = z_c$.

Using conditions (3.15) leads to the following system of equations for the SBS process:

$$\frac{n}{c}\frac{\partial I_L}{\partial t} + \frac{\partial I_L}{\partial z} = -\alpha I_L - g_B I_L I_S \qquad \frac{n}{c}\frac{\partial I_S}{\partial t} - \frac{\partial I_S}{\partial z} = -\alpha I_S + g_B I_L I_S \quad (3.16)$$

where α is the linear optical loss in the material and $g_B = (\omega_s^2(\gamma_e)^2/c^3 n v \rho_0)\tau_B$ is the optical gain associated to the SBS process. This system of equations has been used also in [3.34, 3.37].

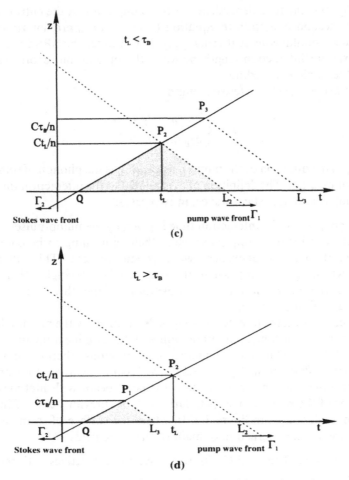

Figure 3.2. (c) The surface (QP_2L_2) is the domain of interaction when $t_L < \tau_B$ and $z_c = ct_L/n$; Γ_1 and Γ_2 are the light wavefronts; Γ_3 is the acoustic wavefront; P_3L_3 is the pump wavefront at $z = c\tau/n$. (d) The surface of triangle (QP_1L_3) represents the domain of interaction when $t_L > \tau_B$ and $z_c = c\tau_B/n$.

To solve analytically this set of nonlinear coupled equations, we shall use the integration on characteristic equations [3.8], a method that leads to more general solutions (analytical distributions in time, t, and space, z, for an arbitrary temporal shape of the optical pump pulse).

In the space (z, t), the characteristic equations associated to system (3.16) have the form (figure 3.2(a)):

$$\xi_L = (c/n)t + z, \qquad \xi_S = (c/n)t + z_c - z \qquad (3.17)$$

The boundary conditions for the system of equations (3.16) are (figure 3.2(b)):

$$I_L(z,t)|_{z=0} = I_{L0}(t), \qquad I_S(z,t)|_{z=z_c} = I_{S0}(t) \qquad (3.18)$$

where $I_{L0}(t)$ is the time dependence of the pump pulse at the entrance of the cell. The second condition in equation (3.18) is an approximation for the distributed spontaneous scattering (I_{S0}) that initiates the SBS process. z_c is the maximum interaction length between the optical and acoustical fields within the nonlinear medium.

We define the characteristic length

$$z_c = \begin{cases} ct_L/n, & t_L < \tau_B \\ c\tau_B/n, & t_L > \tau_B \end{cases} \tag{3.19}$$

where t_L is the duration of the pump pulse and τ_B is the phonon lifetime in the nonlinear medium. The definition of z_c is related to the interaction time of the pump (light) and acoustical waves, in two cases:

(a) When $t_L < \tau_B$, the interaction time is given by the pump-pulse duration, because after the pump pulse ends, there is nothing to be compressed even though the acoustical wave is still present. This situation is shown in figure 3.2(c), where the surface of the triangle QP_2L_2 represents the interaction (superimposition) of the three fields (pump, acoustical and Stokes).

(b) When $t_L > \tau_B$, the interaction time is determined by the phonon lifetime, τ_B, because the long light interaction with the nonlinear medium is more affected by instabilities of the acoustic phonons (thermal and hydrodynamic fluctuations) [3.14, 3.15]. We assume that random processes determine the coherence of the conjugated beam with lifetimes of the order of the phonon ones (similarly with the previous case). This situation is illustrated in figure 3.2(d), where the surface of triangle QP_1L_3 represents the interaction domain of the three fields.

Using equations (3.16) and (3.19), we derive the derivatives with respect to ξ_L and ξ_S:

$$\frac{\partial}{\partial \xi_L} = \frac{1}{2}\left(\frac{n}{c}\frac{\partial}{\partial t} + \frac{\partial}{\partial z}\right), \qquad \frac{\partial}{\partial \xi_S} = \frac{1}{2}\left(\frac{n}{c}\frac{\partial}{\partial t} - \frac{\partial}{\partial z}\right). \tag{3.20}$$

In order to get a simple parametric representation of equations (3.16), we shall write the derivatives as a function of $\partial/\partial \xi_S$ in the form

$$\frac{\partial}{\partial \xi_S} = \frac{1}{2}\left(\frac{n}{c}\frac{\partial}{\partial t} + \frac{\partial}{\partial(z_C - z)}\right) \tag{3.21}$$

and we shall use the integration along the characteristic lines, in a similar manner as Yariv [3.15] has done, with

$$\left|\frac{\partial}{\partial \xi_L}\right| = \left|\frac{\partial}{\partial \xi_S}\right| = \frac{d}{d\eta} \tag{3.22}$$

where η is an integration variable, which is considered along both characteristic lines (figure 3.2(a)). From equations (3.16), (3.20), (3.22) we have

$$\frac{dI_L}{d\eta} = -\alpha I_L - g_B I_L I_S, \qquad \frac{dI_S}{d\eta} = -\alpha I_S + g_B I_L I_S. \qquad (3.23)$$

The solution of (3.23) has the general form

$$I_L\left(\frac{c}{n}t + z\right) = \frac{I_{L0}(t)[I_{L0}(t) + I_{S0}]\exp(-\alpha z)}{I_{L0}(t) + I_{S0}\exp\{g_B[I_{L0}(t) + I_{S0}]\cdot l(z)\}}$$

$$I_{S0}[I_{L0}(t) + I_{S0}] \qquad (3.24)$$

$$I_S\left(\frac{c}{n}t + z_c - z\right) = \frac{\times \exp\{-\alpha(z_c - z) + g_B[I_{L0}(t) + I_{S0}]l(z_c - z)\}}{I_{L0}(t) + I_{S0}\exp\{g_B[I_{L0}(t) + I_{S0}]\cdot l(z_c - z)\}}$$

where $l(z) = [1 - \exp(-\alpha z)]/\alpha$. Taking into account the specific transformation for the backward scattering described in equations (3.22), the optical pump field at the exit of the interaction region, limited by z_c, has the form

$$I_L(t) = I_L\left(\frac{c}{n}t + z\right)\bigg|_{z=z_c} \qquad (3.25)$$

and the Stokes field at $z = 0$ is

$$I_S(t) = I_S\left(\frac{c}{n}t + z_C - z\right)\bigg|_{z=0}. \qquad (3.26)$$

From equations (3.24), one can deduce the intensity of the Stokes field scattered by the cell filled with the nonlinear material (at $z = 0$):

$$I_S(t) = \frac{I_{L0}(t)\exp[-\alpha z_C + g_B I_{L0}(t)\cdot l(z_c) - G_0]}{1 + \exp[g_B I_{L0}(t)\cdot l(z_c) - G_0]} \qquad (3.27)$$

where $G_0 = \ln(I_{L0}/I_{S0})$, I_{L0} is the pump intensity at the entrance of the cell and I_{S0} is the intensity of the spontaneous Stokes field. A particular form of equation (3.27), in the case $\alpha = 0$, was deduced by an alternative method by Johnson and Marburger [3.13] and Chen and Bao [3.16]. In their result, z_c is the interaction length.

From equation (3.27), considering equal cross-sections for the pump and the Stokes waves and different pulse durations, we can find the energy reflectivity of the SBS process (or energy conversion efficiency) as

$$R_{SBS} = \frac{\varepsilon_S}{\varepsilon_L} = \frac{\int_0^{t_S} I_S(t)\,dt}{\int_0^{t_L} I_{L0}(t)\,dt} = \frac{A\,e^{-\alpha z_c}}{\frac{\sqrt{\pi}}{2}\,\text{erf}(1)}\int_0^{t_S/t_L}\frac{e^{-u^2/u_1^2}}{1 + A\,e^{-u^2/u_2^2}}\,du \qquad (3.28)$$

where ε_L is the pump energy, ε_S is the Stokes energy and

$$A = \exp\left(\frac{\varepsilon_L}{\varepsilon_{LI}} - G_0\right), \qquad u_1 = \frac{1}{\sqrt{1 + (\varepsilon_L/\varepsilon_{LI})}}$$

$$u_2 = \frac{1}{\sqrt{\varepsilon_L/\varepsilon_{LI}}}, \qquad u = \frac{t_S}{t_L}, \qquad \frac{\varepsilon_L}{\varepsilon_{LI}} = g_B I_0 z_c$$

(assuming Gaussian pulse shape). We can calculate easier the values of the SBS reflectivity for (1) small pump energy ($\varepsilon_L \to 0$):

$$R_{SBS} \approx \frac{1}{\mathrm{erf}(1)} \exp\left(\frac{\varepsilon_L}{\varepsilon_{LI}} - \frac{2}{\sqrt{\pi}\,\mathrm{erf}(1)} \frac{\varepsilon_{LI}}{\varepsilon_L} - G_0 - \alpha z_c\right)$$

$$\times \left[1 + \exp\left(\frac{\varepsilon_L}{\varepsilon_{LI}} - G_0\right)\right]^{-1} \tag{3.28a}$$

and (2) large pump energy ($\varepsilon_L \to \infty$):

$$R_{SBS} \approx \exp(-\alpha z_c). \tag{3.28b}$$

From the same solution, one can find the intensity of the perturbed pump wave at $z = z_c$:

$$I_L(t) = I_{L0}(t) \exp[-\alpha z_C - g_B I_{L0}(t) \cdot L(z_c) + G_0]. \tag{3.29}$$

3.5   Laser pulse compression by SBS

In order to calculate the duration of the Stokes pulse, one can choose the following form for the optical pump pulse

$$I_{L0}(t) = I_0 f(t), \tag{3.30}$$

where I_0 is the maximum intensity of the optic pump pulse and the envelope function $f(t)$ is continuous together with its derivatives up to the second one, with

$$\begin{aligned} 0 < f(t) \leq 1 & \quad \text{if } t \in [0, t_L] \\ f(t) = 0, & \quad \text{elsewhere} \end{aligned} \tag{3.31}$$

It is assumed that the pump pulse duration is much shorter than the time required for light to traverse the SBS cell.

If the pump pulse is step-like, the scattered pulse will be different. Thus, the scattered pulse has a different shape than the pump pulse, which is dependent on the shape of the pump pulse.

We calculate the duration of the Stokes pulse at $1/e$ from the maximum value of the Stokes intensity, t_S, for the particular shape of the pump pulse:

$$I_{L0} = I_0 \left[2\left(\frac{2t}{t_L}\right) - \left(\frac{2t}{t_L}\right)^2\right]. \tag{3.32}$$

The pulse defined by equation (3.32) is close to commonly used experimental pump pulses and offers the possibility to derive an analytical expression for the Stokes pulse and the compression ratio.

Thus, in the case that the pump pulse is shorter than the phonon lifetime $(t_L < \tau)$, one can obtain

$$t_s = \frac{t_L}{[g_B I_0 \cdot l(z_c)]^{1/2}} \quad \text{for } z_c = \frac{c t_L}{n}. \tag{3.33}$$

The temporal width obtained by Hon [3.3, 3.17] for the scattered pulse is a particular case of that of equation (3.33), when $\alpha = 0$.

From equations (3.27) and (3.33), we can obtain the compression ratio (t_L/t_S) [3.18–3.20]:

$$\left(\frac{t_L}{t_S}\right)_{t_L < \tau} = \begin{cases} [g_B I_0 \cdot l(z_c)]^{1/2}, & \text{for } g_B I_{L0} \cdot l(z_c) - G_0 < 0 \\ 1, & \text{for } g_B I_{L0} \cdot l(z_c) - G_0 \gg 0. \end{cases} \tag{3.34}$$

Similar to equation (3.34), equation(3.27) may be written in the form

$$\left(\frac{I_S(t)}{I_{L0}(t)}\right)_{t_L < \tau}$$

$$= \begin{cases} \exp[-\alpha z_c + g_B I_{L0}(t) \cdot l(z_c) - G], & \text{for } g_B I_{L0}(t) \cdot l(z_c) - G_0 < 0 \\ \exp(-\alpha z_c), & \text{for } g_B I_{L0}(t) \cdot l(z_c) - G_0 \gg 0. \end{cases} \tag{3.35}$$

From equations (3.34) and (3.35), one can observe that for small pump intensities the Stokes intensity is proportional to the pump intensity and the compression ratio has high values.

For high pump intensities, a saturation process appears for the Stokes intensity and the compression ratio becomes unity. In this region, there is no compression $(t_L/t_S \to 1)$ and the Stokes intensity is

$$I_S(t) = I_{L0}(t) \exp(-\alpha z_c) \qquad (t_L < \tau_B). \tag{3.36}$$

We suggested calling this regime a 'quasi-stationary regime', as an extension of the definition given by Kaiser and Maier [3.1] and Kroll [3.2] and commonly used in the SBS literature [3.18–3.20]. In this regime, the temporal dependences of the scattered and pump pulses are similar up to the multiplicative constant $[\exp(-\alpha c t L/n)]$.

When the duration of the pump pulse is longer than the phonon lifetime $(t_L > \tau_B)$ and the pump intensity is high, using the expression for z_c from equation (3.19), we obtain

$$I_S(t) = \frac{I_{L0}(t) \exp\left[-\alpha \frac{c\tau_B}{n} + g_B I_{L0}(t) \cdot l\left(\frac{c\tau_B}{n}\right) - G_0\right]}{1 + \exp\left[g_B I_{L0}(t) \cdot l\left(\frac{c\tau_B}{n}\right) - G_0\right]}, \tag{3.37}$$

a similar expression to that of equation (3.27).

Taking the similar shape for the pump pulse as that in equation (3.32), the duration of the Stokes pulse becomes

$$t_s = \frac{t_L}{[g_B I_0 \cdot l(c\tau_B/n)]^{1/2}}. \tag{3.38}$$

In this case, the compression ratio, (equivalent to equation (3.34)) is:

$$\left(\frac{t_L}{t_S}\right)_{t_L > \tau} = \begin{cases} [g_B I_0 \cdot l(c\tau_B/n)]^{1/2}, & \text{for } g_B I_{L0} \cdot l(c\tau_B/n) - G_0 < 0 \\ 1, & \text{for } g_B I_{L0} \cdot l(c\tau_B/n) - G_0 > 0 \end{cases}. \tag{3.39}$$

In both cases, $t_L > \tau_B$ or $t_L < \tau_B$, we have defined a quasi steady-state regime of SBS in the saturation region of our general solutions (3.27) and (3.37), which are dependent on the pump pulse intensity and duration, and on the absorption of the nonlinear medium. The usual definition of this regime, implying the cancellation of the temporal derivatives [3.1, 3.2] leads, in our opinion, to an over-simplification of the process evolution and to an insufficient condition for the existence of such steady-state regime.

From equations (3.34) and (3.39), we can deduce the condition for pulse compression ($t_L/t_S > 1$). In the case $t_L > \tau_B$ and $\alpha = 0$, this condition is

$$G_0 \geq g_B I_0 \frac{c\tau_B}{n}. \tag{3.40}$$

In the case $t_L < \tau_B$ and $\alpha = 0$, the pulse compression appears when

$$G_0 \geq g_B^e I_0 \frac{ct_L}{n}. \tag{3.41}$$

One can remark that the compression conditions obtained in equations (3.40) and (3.41), are similar up to the times involved, τ_B and t_L respectively. They are valid only for pump pulses as defined in equation (3.32).

3.6   Stochastic processes in the solution of SBS equations

When the duration of the pump pulse is longer than the phonon lifetime, $t_L > \tau_B$, and the pump intensity is high, the scattering will take place on the fluid disturbances with random character (great fluctuations of the thermal field, induced turbulence, multiplicative ionisation phenomena, shock hydrodynamic waves, etc.), besides the Brillouin scattering [3.14, 3.15, 3.17, 3.20–3.22]. The equations for the SBS process could be written under the form [3.18]

$$\frac{n}{c}\frac{\partial I_L}{\partial t} + \frac{\partial I_L}{\partial z} = -\alpha I_L - (1 - \varepsilon') \cdot g_B I_L I_S$$

$$\frac{n}{c}\frac{\partial I_S}{\partial t} - \frac{\partial I_S}{\partial z} = -\alpha I_S + (1 - \varepsilon') \cdot g_B I_L I_S$$

(3.42)

where ε' describes the random phase fluctuations related to SBS medium density fluctuations (noise).

In the simplest case, the random variable ε' has a Gaussian distribution, with zero mean and dispersion σ:

$$\langle \varepsilon'(\eta) \rangle = 0, \qquad \langle \varepsilon'(\eta) \cdot \varepsilon'(\eta') \rangle = 2\sigma_d^2 g_B I_0 \delta(\eta - \eta') \qquad (3.43)$$

where $\langle \cdots \rangle$ means the average, $\delta(\xi - \xi')$ is the Dirac function and σ_d is the dispersion of the Gaussian process which depends on the properties of the SBS medium. In the classical theory of the optical coherence, the degree of coherence has, from the mathematical point of view, the significance of the dispersion of the stochastic process.

In our case, the square of the dispersion (σ_d^2) of the stochastic process can be interpreted as an effective interaction length, which can be smaller or higher than the maximum coherence length $(c\tau_B/n)$.

In this case, the phase condition (3.13) takes the form

$$\varepsilon' = \varphi_L - \varphi_S - \varphi_d \qquad (3.44)$$

with φ_d as the disturbance process.

The rigorous analysis (the solution of the Navier–Stokes equation in the complete form) of the influence of all factors that disturb the SBS process is very difficult. We shall use the method of separation of the slow variables against the rapid ones (related to the random Gaussian disturbing process mentioned before).

The steps in this mathematical procedure [3.18–3.22] pass through the building of the statistical Liouville equation (SLE) in the intensity space, then the deduction of Fokker–Planck–Kolmogorov equation (by SLE averaging), and finally the deduction of the evolution equations for the mean values of the intensities, I_L and I_S. The process of averaging equations (3.42) is shown in appendix 1.

In this manner, one can obtain a set of deterministic equations that describes the evolution of the mean values of the intensities on the characteristic lines:

$$\frac{n}{c}\frac{\partial I_L}{\partial t} + \frac{\partial I_L}{\partial z} = -\alpha I_L - g_B I_L I_S - \sigma_d^2 g_B^2 \cdot (I_L I_S^2 - I_L^2 I_S)$$

$$\frac{n}{c}\frac{\partial I_S}{\partial t} - \frac{\partial I_S}{\partial z} = -\alpha I_S + g_B I_L I_S - \sigma_d^2 g_B^2 \cdot (I_L^2 I_S - I_L I_S^2).$$

(3.45)

Using the boundary conditions (3.18), one can arrive at the following prime integrals for the pump and the Stokes wave intensities [3.18]:

$$I_L\left|I_L - \frac{1}{2}\left(I_{L0} + \frac{1}{\sigma_d^2 g_B}\right)\right|^{(2\sigma_d^2 g_B I_{L0})/(1-\sigma_d^2 g_B I_{L0})}$$

$$= c_2|I_L - I_{L0}|^{(1+\sigma_d^2 g_B I_{L0})/(1-\sigma_d^2 g_B I_{L0})} \exp[-(\sigma_d^2 g_B^2 I_{L0}^2 + g_B I_{L0})\eta]$$

$$I_S = c_3|I_S - I_{L0}|^{(1-\sigma_d^2 g_B I_{L0})/(1+\sigma_d^2 g_B I_{L0})} \tag{3.46}$$

$$\cdot \left|I_S - \frac{1}{2}\left(I_{L0} - \frac{1}{\sigma_d^2 g_B}\right)\right|^{(2\sigma_d^2 g_B I_{L0})/(1+\sigma_d^2 g_B I_{L0})} \exp[-(\sigma_d^2 g_B^2 I_{L0}^2 - g_B I_{L0})\eta]$$

where c_2 and c_3 are integration constants.

In the deterministic limit, $\sigma_d^2 \to 0$, equations (3.46) are identical to equations (3.27) and (3.29), derived by us for SBS without noise.

For large optical gain and for small noise dispersion, solutions (3.46) become

$$I_L(t) = \frac{I_{L0}(t)}{1 + \exp[(g_B I_{L0} - \sigma_d^2 g_B I_{L0}^2)(c\tau/n) - G_0]}$$

$$I_S(t) = \frac{I_{L0}(t)\exp[(g_B I_{L0} - \alpha - \sigma_d^2 g_B I_{L0}^2)(c\tau/n) - G_0]}{1 + \exp[(g_B I_{L0} - \sigma_d^2 g_B I_{L0}^2)(c\tau/n) - G_0]}. \tag{3.47}$$

We notice that solutions (3.47) differ from equations (3.27) and (3.29) by a supplementary term in the exponential gain $(\sigma_d^2 g_B^2 I_L 0^2)$, which can be interpreted as an additional diffusion process of the optical field. Similarly, we can calculate a compression ratio (for $t_L > \tau_B$)

$$\frac{t_L}{t_s} = \left[g_B I_0 \cdot l\left(\frac{c\tau_B}{n}\right)\right]^{1/2} \cdot \left\{\frac{1}{2}\left[(1 - 2\sigma_d^2 g_B I_{L0})^2 + 4\frac{\sigma_d^2 n}{c\tau}\right]^{1/2} + 1 - 2\sigma_d^2 g_B I_0\right\}^{1/2}. \tag{3.48}$$

In equation (3.48), we notice that the limit of the compression ratio for $t_L > \tau_B$, as $\sigma_d^2 \to 0$, equals the compression ratio for $t_L > \tau_B$

$$\lim_{\sigma_d^2 \to 0} \left(\frac{t_L}{t_s}\right)\bigg|_{t_L > \tau_B} = \left(\frac{t_L}{t_s}\right)\bigg|_{t_L > \tau_B}. \tag{3.49}$$

Thus, the statistical modelling of SBS leads, in the limit $\sigma_d^2 \to 0$, to the same results as the deterministic modelling.

3.7 The experimental verification of the analytical results in the one-dimensional SBS model

The theoretical results above have been checked experimentally in a conventional configuration for SBS using a Nd:YAG laser (oscillator–amplifier

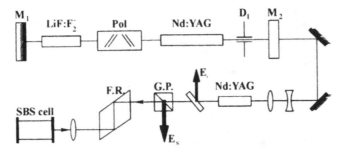

Figure 3.3. The experimental set-up for SBS analytical result verification.

system) and carbon disulphide in a glass cell as the nonlinear material (figure 3.3).

The laser oscillator was operated in the Q-switched mode using a LiF:F$_2$ crystal with an initial transmission of 7%. The oscillator was kept near threshold in order to generate a single Q-switch pulse with the flashlamp–pump energy of 15 J. The resonator length was 40 cm. The short resonator and the Q-switch with a small initial transmission allowed the generation of a pulse of 8 ns duration. A stack of four glass plates (Brewster angle) was used to get a linearly polarized output. To achieve the transverse mode selection, an internal aperture of 1.5 mm diameter was used.

The output pulse energy of the oscillator was amplified, by a single pass, in a second Nd:YAG module. The output pulse energy of the oscillator–amplifier laser system was 40 mJ.

The amplified impulse was focused with a convergent lens onto a cell containing CS$_2$ as the nonlinear SBS medium. Between the amplifier and the cell, an optical isolator (with a Glan prism and a Fresnel rhombus) was introduced. The first pass of the Fresnel rhombus resulted in the linear polarized light being converted to circular polarized light. After SBS reflection, the second passage of the rhombus converted the light to linear but orthogonal polarization to the original and this was out-coupled by the polarizer (GP). This decoupling system prevented feedback into the laser and also allowed ease of monitoring of the SBS return.

The energy incident on the cell and the energy backscattered from the cell were measured with a calorimeter. The temporal laser beam evolution was recorded by a fast photodiode and a high bandwidth oscilloscope.

With the experimental system presented in figure 3.3, it was possible to make systematic pulse compression measurements in the Brillouin scattering, for $t_L > \tau_B$.

For CS$_2$, the parameters $g_B = 0.06$ cm/MW, $\tau_B = 6$ ns, at $\lambda = 1.06$ µm [3.23] were used. The SBS cell used in this experiment was long enough (1.2 m) in order to allow the interaction length given by equation (3.19), when the laser pump pulse width is larger than the phonon lifetime.

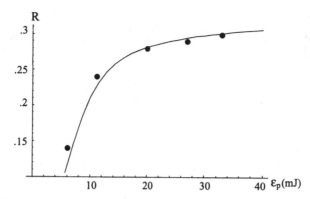

Figure 3.4. Dependence of the energy reflectivity $(\varepsilon_S/\varepsilon_L)$ on the pump energy, for $t_L = 8\,\text{ns}$. The theoretical curve is drawn with continuous line, using equation (3.28a). Points represent the experimental data.

In these experiments, a long focal length of the lens (1 m) was used in order to have a reduced focusing and be able to compare the experimental results with the calculations done for a plane wave.

The dependence of the SBS energy reflectivity $(\varepsilon_S/\varepsilon_L)$ on the pump energy, for the pump pulse duration of $t_L = 8\,\text{ns}$ is shown in figure 3.4. We have considered the pump intensity as $I_{L0} = \varepsilon_L/[\pi(d_L/2)^2 t_L]$, with $d_L = 0.4\,\text{cm}$ the diameter of the incident laser pulse and the laser pulse compression negligible. The stationary regime, defined by equations (3.15), appears as the lines on these graphs for pump energies less than and larger than 10 mJ. A good fit of the experimental data with equation (3.28) is found.

In the transient regime, the compression ratio (t_L/t_S) is small (maximum 8, for a pump energy of 40 mJ), increasing and saturating with the pump energy.

The dependence of the pulse–compression ratio against the pump energy (derived in the stochastic formalism) is illustrated in figure 3.5, for different values of the laser pulse duration and of the noise dispersion, σ_d. The dependence found by Hon (for $\alpha = 0$ and $\sigma_d = 0$) [3.3] is equally shown, in order to compare the theoretical predictions and the importance of noise consideration. We notice that the experimental data are well fitted by the curve corresponding to $\alpha = 0.028\,\text{cm}^{-1}$ (measured experimentally) and to the noise dispersion $\sigma_d = 0.3$, which depends on the properties of the SBS medium (which were taken, in our case, without a complete characterization and purification). One can remark that the stochastic theory is valid even in our experimental conditions, when the duration of the incident laser pulse is close to the lifetime of the acoustic phonons, τ_B.

The remaining discrepancies between the theoretical prediction of equation (3.46) and the experimental data can be explained by the small

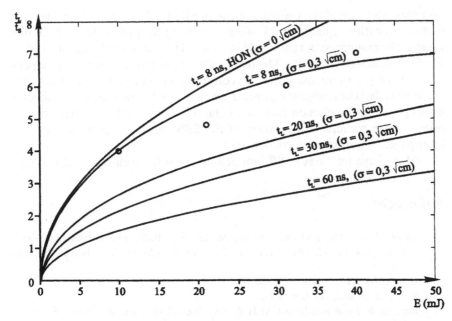

Figure 3.5. Dependence of the compression ratio (t_L/t_S) on optical pump energy, for different values of the laser pulse duration. The theoretical curves (continuous lines) were derived from equation (3.46) and the experimental points are represented by circles.

laser pump duration used in this experiment (8 ns) which was not much larger compared with the phonon lifetime in CS_2 (6 ns).

3.8 Conclusions

In this chapter, we have considered that the SBS process involves an incident pump beam with a plane wavefront and the scattered Stokes wave is also considered to have a plane wavefront. In this regime, the SBS process can be considered as a one-dimensional interaction.

Analytical and numerical solutions of the SBS system of equations for steady-state and transient cases have been presented.

Solving the SBS equations by the method of characteristic equations, accurate analytical solutions for pump and Stokes intensities were obtained. In this approach, the expressions for the Stokes and pump intensities were deduced for an arbitrary time dependence of the incident pump intensity. The theoretical predictions for SBS reflectivity and for the laser pulse compression show good agreement with the experimental data, which are obtained for pump pulses longer than the phonon lifetime (in carbon disulphide, as the nonlinear medium). The experimental results were better fitted by the stochastic theory, in which we have considered absorption and the dispersion of the Gaussian noise in SBS [3.18, 3.20]. This model predicts a saturation regime of SBS, when increasing the pump intensity. In this regime, the compression ratio tends to unity. Thus, it could be considered as the SBS steady-state regime. The region where the Stokes intensity is dependent on the pump intensity and the compression ratio is higher than unity could be called the transient regime, regardless of whether the pulse is shorter or longer than the phonon lifetime of the nonlinear medium. These results offer a more complete explanation of the SBS process in general and SBS laser pulse compression in particular.

Many investigations in SBS modelling are still developing [3.24–3.37].

References

3.1 Kaiser W and Maier M 1972 Stimulated Rayleigh, Brillouin and Raman spectroscopy in *Laser Handbook* vol 2 ed F T Arecchi (Amsterdam: North-Holland) p 1077

3.2 Kroll N M 1965 *J. Appl. Phys.* **36** 34

3.3 Hon D T 1980 *Optics Lett.* **5** 516

3.4 Damzen M J and Hutchinson M H R 1983 *IEEE J. Quantum Electron.* **QE-19** 7

3.5 Filippo A A and Perrone M R 1992 *J. Mod Opt.* **39** 1829

3.6 Nassisi V and Pecoraro A 1993 *IEEE J. Quantum Electron.* **QE-29** 2547

3.7 Brent C B, Neuman W A and Hackel L A 1994 *IEEE J. Quantum Electron.* **QE-30** 1907

3.8 Rojdestvensky B L and Ivanenko N N 1978 *Quasilinear Systems of Equations* (Moscow: Mir) p 16 (in Russian)

3.9 Armstrong J A, Bloembergen N, Ducuing J and Pershan P S 1962 *Phys. Rev.* **127** 1918

3.10 Maier M 1968 *Phys. Rev.* **166** 113

3.11 Pohl D and Kaiser W 1970 *Phys. Rev. B* **1** 31

3.12 Kroll N M and Kelley P L 1971 *Phys. Rev. A* **4** 763

3.13 Johnson R V and Marburger J H 1971 *Phys. Rev. A* **4** 1175

3.14 Nguyen-Vo N M and Pfeifer S J 1993 *IEEE J. Quantum Electron.* **QE-29** 508

3.15 Yariv 1975 *Quantum Electronics* 2nd edition (New York: Wiley) p 387

3.16 Chen L and Bao X 1998 *Opt. Commun.* **152** 65

3.17 Hon D T 1982 *Opt. Eng.* **21** 252

3.18 Babin V, Mocofanescu A, Vlad V and Damzen M J 1999 *J. Opt. Soc. Am. B* **16** 155; 1995 *Proc. SPIE* **2461** 294

3.19 Vlad V I, Babin V and Mocofanescu A 1995 *Proc. SPIE* **2461** 294

3.20 Babin V, Mocofanescu A and Vlad V I 1996 *Ro. Repts. Phys.* **3-4** 299

3.21 Kliatkin V I 1981 *Inst. Appl. Phys. USSR Academy, Moscow* (in Russian)

3.22 Kliatkin V I 1986 *Averaging Problems and the Scattering Theory of Waves* (Moscow: Mir) (in Russian)

3.23 Erokhin A, Kovalev V I and Faizullov S F 1986 *Sov. J. Quantum Electron.* **16** 872

3.24 Tikhonchuk V T, Labaune C and Baldis H A 1996 *Phys. Plasmas* **3** 3777

3.25 Giacone R E and Vu H X 1998 *Phys. Plasmas* **5** 1455

3.26 Labaune C, Baldis H A and Tikhonchuk V T 1997 *Europhysics Lett.* **38** 31

3.27 Klovekorn P and Munch J 1997 *Appl. Opt.* **36** 5913

3.28 Jo M S and Nam C H 1997 *Appl. Opt.* **36** 1149

3.29 Sirazetdinov V S and Charukhchev Av 1997 *Opt. Tech.* **64** 1151

3.30 Yashin V E, Chizhov S A, Gorbunov V A and Lavrent K K 1998 *Proc. SPIE* **3264** 43

3.31 Andreev N F, Palashov O V and Khazanov E A 1999 *Quantum Electron.* **29** 314

3.32 Neshev D, Velchev I, Majewski W A, Hogervorst W and Ubachs W 1999 *Appl. Phys. B* **68** 671

3.33 Velchev I, Neshev D, Hogervorst W and Ubachs W 1999 *IEEE J. Quantum Electron.* **35** 1812

3.34 Lecoeuche V, Segard B and Zemmouri J 1999 *Opt. Commun.* **172** 335

3.35 Vlad V I, Damzen M J, Babin V and Mocofanescu A 2000 *Stimulated Brillouin Scattering* (Bucharest: INOE Press)

3.36 Boyd R W, Rzazewski K and Narum P 1990 *Phys. Rev. A* **42** 5514

3.37 Montes C, Bahloul D, Bongrand I, Botineau J, Cheval G, Mamhoud A, Picholle E and Picozzi A 1999 *J. Opt. Soc. Am. B* **16** 932

Chapter 4

Optical phase conjugation in SBS

4.1   Phase conjugation and aberration compensation

Nonlinear optical phase conjugation has been shown to have a unique ability to restore an aberrated beam to its original undistorted state in real time [4.1–4.11]. Independent of the precise physical method of production, the important feature of the phase conjugation process is the creation of a wave whose optical field amplitude $A_C(\mathbf{r})$ is proportional to the complex conjugate of an input field $A_i(\mathbf{r})$, i.e. $A_C(\mathbf{r}) \propto A_i^*(\mathbf{r})$.

This behaviour can be visualized by considering an input wave with optical field $E(\mathbf{r}, t)$ represented in the form

$$E(\mathbf{r}, t) = \tfrac{1}{2}[E(\mathbf{r}) \exp i(\omega t - \mathbf{k} \cdot \mathbf{r}) + \text{c.c.}] \tag{4.1a}$$

or alternatively

$$E(\mathbf{r}, t) = \tfrac{1}{2}[|E(\mathbf{r})| \exp i(\omega t - \mathbf{k} \cdot \mathbf{r} + \phi(\mathbf{r})) + \text{c.c.}] \tag{4.1b}$$

where $E(\mathbf{r}) = |E(\mathbf{r})| \exp(i\phi(\mathbf{r}))$ is the complex slowly-varying amplitude of the electric field, $\phi(\mathbf{r})$ is the phase of the wave, and ω and $\mathbf{k}$ are the angular frequency and wavevector, respectively. For a plane wave (with plane wavefronts) $\phi(\mathbf{r})$ is a constant, whereas for non-plane waves (e.g. with aberrated wavefronts) phase fronts are more complicated as depicted in figure 4.1.

In this notation, the phase conjugate wave is given by

$$E_c(\mathbf{r}, t) = \tfrac{1}{2}r_c[|E(\mathbf{r})| \exp i(\omega t + \mathbf{k} \cdot \mathbf{r} - \phi(\mathbf{r})) + \text{c.c.}] \tag{4.2}$$

where the wavevector is reversed to the incident wave ($\mathbf{k}_c = -\mathbf{k}$) such that it counter-propagates to the input field and the phase front is reversed ($\phi_c(\mathbf{r}) = -\phi(\mathbf{r})$), i.e. the field is the complex conjugate of the incident wave. The quantity r_c is the proportionality constant and can be considered the (amplitude) reflectivity of the phase conjugation process. The reflectivity r_c is, in general, a complex quantity although its absolute phase is not usually important except in interferometric applications.

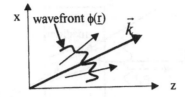

Figure 4.1. An optical field with aberrated wavefront.

The key importance of the phase conjugation technique is its ability to correct for phase aberrations encountered by the incident beam by double-passing the distorting beam path with the phase conjugate beam, restoring the beam to its original quality. A nonlinear optical device that produces the phase conjugate reflection is known as a phase conjugate mirror (PCM). This correction process is depicted in figure 4.2.

The most commonly used methods to achieve phase conjugation are four-wave mixing (FWM) [4.12] and stimulated Brillouin scattering (SBS) [4.13] as shown in figure 4.3(a) and (b), respectively. Four-wave mixing (FWM) involves the input of three beams into a nonlinear medium with third-order susceptibility $\chi^{(3)}$. Two of the beams, E_1 and E_2 (called pump beams), are usually counter-propagating and the other beam E_3 (signal beam) is incident at an angle. The interaction generates a fourth beam E_4 (conjugate beam) that is the phase conjugate to E_3. Stimulated Brillouin scattering (SBS) involves nonlinear formation of an acoustic wave 'mirror'

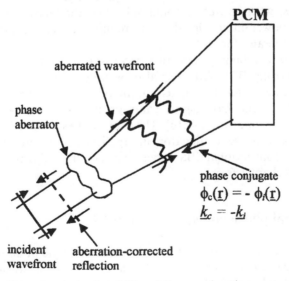

Figure 4.2. The aberration-correcting ability of phase conjugation.

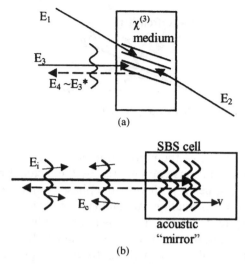

Figure 4.3. Phase conjugation produced by (a) four-wave mixing in a nonlinear medium and (b) stimulated Brillouin scattering.

due to an intense input beam. The maximum nonlinear interaction occurs when the reflected beam is a phase conjugate of the input beam. SBS is an example of a self-conjugating process in that no other input beams or external controls are necessary to produce the phase conjugation.

These techniques differ widely in their input requirements and as such are usually operated in different regimes. FWM can be used for phase conjugation of a weak signal beam but requires high power and high quality pump beams, whereas SBS is commonly used for phase conjugation of a single high power input signal, as it requires a certain threshold input intensity to operate efficiently.

A system combining these two processes, termed Brillouin-enhanced four-wave mixing (BEFWM), has also been demonstrated. This combines the desirable characteristics of its parent processes in that it has no SBS threshold [4.14], can have signal gains greater than 10^5 [4.15] and gives excellent phase conjugation [4.16]. These properties make BEFWM a good candidate as a phase conjugator in applications where the low input powers make conventional SBS impossible. The BEFWM process is treated separately in chapter 6.

4.2 Optical phase conjugation by SBS

Optical phase conjugation by SBS, and indeed the topic of nonlinear optical phase conjugation, was brought to the attention of the world in 1972 by Zel'dovich *et al* [4.1, 4.2]. By focusing an aberrated ruby laser beam into a

1 m long multimode lightguide filled with methane gas at a pressure of 125 atm they generated a backscattered Stokes wave. After a second pass through the aberration the backward Stokes wave compensated the aberrations and reproduced the unaberrated wavefront of the incident laser beam.

Phase conjugation by SBS can often work better the more aberrated the pump. Zel'dovich *et al* [4.2] were the first to offer an explanation of the phenomenon and theory further advanced by contributions of Sidorovich [4.3] and Yariv [4.4]. Neglecting pump depletion, the propagation of pump wave and growth of the Stokes wave is governed by

$$\frac{\partial E_L}{\partial z} + \mathrm{i}\frac{1}{2k}\nabla_\perp E_L = 0, \qquad \frac{\partial E_S}{\partial z} - \mathrm{i}\frac{1}{2k}\nabla_\perp E_S = -\frac{1}{2}g_B|E_L|^2 E_S \qquad (4.4)$$

assuming the field $E(r, z)$ of the pump and Stokes waves nonuniform at position z and r is a vector coordinate transverse to the direction of propagation (the z direction), and defining field units as $E(r, z) = I(r, z)^{1/2}$, where $I(r, z)$ is the spatial intensity distribution. A useful approach is to express the growth in the Stokes power P_S by integrating the field equation (4.4) over the transverse coordinate r and using relation $P_S(z) = \int \int |E_S(r, z)|^2 \, \mathrm{d}^2 r$ to give

$$\frac{\mathrm{d}P_S(z)}{\mathrm{d}z} = G(z) \cdot P_S(z) \qquad (4.5)$$

where an effective Brillouin-gain factor can be written as

$$G(z) = g_B \frac{\displaystyle\int |E_L(r, z)|^2 |E_S(r, z)|^2 \, \mathrm{d}^2 r}{\displaystyle\int |E_S(r, z)|^2 \, \mathrm{d}^2 r}. \qquad (4.6)$$

In the above derivation, integration of the transverse derivative term over the transverse boundary $\int E_S^* \nabla_T E_s + E_s \nabla_T E_S^* \, \mathrm{d}^2 r$ has been taken as zero since the field and its transverse derivative are expected to tend to zero at large transverse distance for physical beam distributions.

In the case when the two field distributions are phase conjugate $E_S(r, z) = r_c E_L^*(r, z)$, the distributions are perfectly correlated and the Stokes backscattered wave is the phase conjugate replica of the incident wave. The effective Brillouin-gain intensity factor is maximum. For fields with Gaussian statistics, the relationship of equation (4.6), using central limit theorem $\langle x^n \rangle = n!\langle x \rangle^n$, where $\langle \ \rangle$ is the spatial average, gives $G(z) = 2g_B\langle I_L(z)\rangle$. When the two beams are uncorrelated $G(z) = g_B\langle I_L(z)\rangle$. We can say that the Brillouin backscattered wave which is the phase conjugate replica of the incident wave represents the field configuration with the highest spatial gain. This enhanced gain is the reason why it is favoured in the backscattering process. It is demonstrated that the gain factor for the wave at maximum spatial overlap (i.e. the conjugate wave) is about a factor of 2 larger than that for an uncorrelated scattered wave [4.3, 4.4]. The above analysis is only strictly valid for an interaction geometry that is bounded in the transverse dimensions. This case applies to a waveguide interaction. This topic will be analysed in more detail in chapter 8, relating to SBS in optical fibres [4.17–4.35].

A physical picture of this enhanced gain can be described by considering that the backscattered beam will experience most gain if its intensity 'hot-spots' match those of the pump beam. Due to diffraction these hot-spots redistribute in space as the beam propagates. The phase conjugate beam is the only spatial profile that can match its hot-spots with the pump throughout the interaction length.

Consider interaction of SBS in a waveguide structure. A laser intensity I_L is launched into the waveguide and we assume no depletion of the light in the SBS interaction (the small-signal approximation). The Stokes scattering starts from a spontaneous source that occupies the full range of allowed modes of the guide structure. The phase conjugate component of this source has one unique modal distribution with intensity I_{PC} and the non-conjugate component has intensity I_{NPC} with intensity N times the phase conjugate (assume N effective modes of guide). The phase conjugate distribution sees twice the gain coefficient as the uncorrelated modes distributions and the resultant SBS amplified scattered intensity over a guide length (l) is given by

$$I_S = I_{PC} \exp(2g_B I_L l) + I_{NPC} \exp(g_B I_L l). \tag{4.7}$$

We have noted that the enhanced gain leads to selection of the phase conjugate. We can be more quantitative by defining the phase conjugate fraction (H) for the phase conjugate and non-phase conjugate output intensity contributions I'_{PC} and I'_{NPC} as

$$H = \frac{I'_{PC}}{I'_{PC} + I'_{NPC}} = \frac{I_{PC} \exp(2g_B I_L l)}{I_{PC} \exp(2g_B I_L l) + I_{NPC} \exp(g_B I_L l)}. \tag{4.8}$$

As the gain factor $g_B I_L l$ increases, so the value of H approaches unity. For high phase conjugate fidelity $H \approx 1$, we can use expression

$$H = 1 - \frac{I_{NPC}}{I_{PC}} \exp -(g_B I_L l). \tag{4.9}$$

In a light-guide of area A, the number of Stokes modes (N) amplified in an effective angular distribution θ is approximately

$$N = A \frac{\theta^2}{\lambda^2} \approx \frac{I_{NPC}}{I_{PC}}. \tag{4.10}$$

For a typical experimental SBS case: light-pipe diameter $d = 2\,\text{mm}$, $\lambda = 500\,\text{nm}$, $A/\lambda^2 \approx 10^7$. To achieve a fidelity $H > 90\%$, assuming near threshold gain factor $g_B I_L l \approx 15$, requires acceptance angle of guide $\theta < 1$ radian, which is readily achieved. In original experiment by Zel'dovich [4.2], $H = 1$ was measured within experimental errors.

There have been attempts to apply analysis to other interactions geometries, notably the case of a focused speckle beam, comprising a Gaussian distribution of speckles and a Gaussian intensity envelope. By evaluating the overlap integral of equation (4.6), a gain enhancement of the phase conjugate is observed relative to the uncorrelated field.

4.3 Experimental measurement of quality of phase conjugation

Several methods are used to assess beam quality in both a qualitative and quantitative manner. Most of these methods consider properties such as visual appearance in near and far-field or divergence, and in recent ISO standard method assessment of the beam quality factor known as the M^2 of the beam. The determination of whether a beam is phase conjugate can be more problematic as in principle a precise comparison of wavefronts and amplitude distributions needs to be assessed.

A mathematical formulation of the quality of phase conjugation, also known as the phase conjugate fidelity (PCF), is the quantity H defined as

$$H = \frac{\left| \int E_L(r) E_S(r) \, \mathrm{d}^2 r \right|^2}{\int |E_L(r)|^2 \, \mathrm{d}^2 r \int |E_S(r)|^2 \, \mathrm{d}^2 r}. \tag{4.11}$$

The numerator is an overlap integral that provides the degree of correlation of the two fields. It is maximized when E_S is proportional to E_L^*, i.e. wavefront reversal and matching amplitude distribution. The denominator provides normalization to the average powers of the two fields such that H has a maximum value of unity when E_L and E_S are perfect phase conjugates and zero when they are uncorrelated. Experimental measurement of H requires a precise wavefront interferometric technique for providing the PCF. In practice this may not be available and simpler methods have been performed.

4.3.1 Visual assessment and angular spectrum techniques

If the incident beam contains an image, the phase conjugate beam should reproduce the image distribution. For a complex distorted laser distribution, which may contain a speckle distribution, this visual decision may be harder to assess. Even though an amplitude reproduction is suggestive of phase conjugation it does not ensure wavefront reversal. For a fuller assessment one

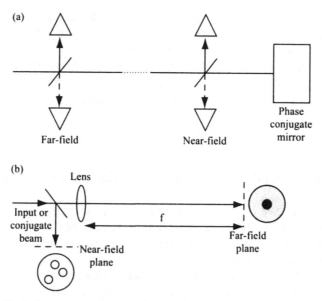

Figure 4.4. Optical systems to assess the quality of phase conjugation using visual observation on CCD cameras (or power meters with pinholes) in (a) near-field and far-field planes of SBS cell and (b) where far-field is obtained in focal plane of lens.

can look in more than one plane. Two obvious planes for observation are the near-field and the far-field planes. The far-field can be reached by observing at a long distance or, more conveniently, at the focal plane of a lens (figure 4.4).

The distribution in the focal plane is the Fourier transform of the near-field distribution. This transformation has some well-known properties that allow some overall features of a beam to be assessed quantitatively. Consider the case when the beam is a perfect Gaussian distribution corresponding to a TEM_{00} laser mode. The Fourier transform of the beam is also a Gaussian distribution but corresponding to the angular spectrum of the near-field distribution. The Fourier transform can be thought of as a reciprocal space in which large spatial features map to small scale features and vice-versa, or can be considered that large features have low spatial frequencies and small features large spatial frequencies. If the Gaussian beam has a noise component on it with small scale features (figure 4.4), the Gaussian component will be transformed to a small Gaussian distribution in the Fourier plane of the lens; the noise features will occupy a larger distribution. A semi-quantitative assessment of the ratio of noise to Gaussian component is found by comparing the power in the total distribution (P_T) to the power passed by a small pinhole with a size that transmits predominantly only the Gaussian component (P_P).

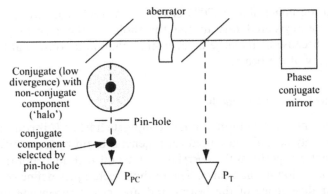

Figure 4.5. Optical system to assess the quality of phase conjugation using high divergence phase plate to enhance discrimination of phase conjugate component.

If the original incident beam was a Gaussian beam, a simple estimate of the phase conjugate fidelity is

$$H = \frac{P_P}{P_T}. \qquad (4.12)$$

This assessment of phase conjugate fidelity is relatively simple to perform but has certain difficulties. First, some of the noise component is transmitted by the pinhole and some of the spatial wings of the Gaussian signal are blocked by the pinhole. A choice needs to be made about the appropriate size of pinhole. There is considerable uncertainty about the value of H if it has a value that is very sensitive to pinhole size. This method works well if the noise (non-phase conjugate) component has a much larger angular spectrum than the wanted (phase conjugate) signal in which case an over-sized pinhole can be used.

A good method of discriminating between the conjugate and non-conjugate components is by placing a strongly distorting phase plate in the path of the incident beam, as shown in figure 4.5. The phase conjugate beam will pass back through the phase plate and be recovered to its original quality, which we assume here is a good quality Gaussian beam of low (diffraction-limited) divergence. The non-phase conjugate component of the reflected beam will not be corrected but will be strongly increased in divergence by a second passage through the phase plate. The pinhole technique will be a good method of assessment of the phase conjugate fidelity (H) due to the high divergence discrimination of phase conjugate and non-phase conjugate components allowing use of over-sized pinhole.

An alternative to using a power meter is to use a photographic film or CCD camera to assess H. These methods allow a good visualization of the spatial form of the beam and qualitative assessment of the quality of phase conjugation. Care needs to be taken to use the appropriate dynamic range of the photographic film or CCD elements. A weak high divergence 'halo' may not be registered due to the threshold of the recording medium, although, because of its large area, it may contain a considerable fraction of the energy of the beam.

4.3.2   Interferometric methods

The most complete definition of phase conjugate fidelity H is equation (4.11) and it can only be truly measured for a general beam by an interferometric method that can perform the correlation integral with incorporation of the phase information in the beams. The pinhole techniques define comparison of angular divergence of the beams and are most appropriate only when the incident beam is a near diffraction-limited beam—which is often the case in laser applications in which aberration correction is required to maintain beam quality.

Figure 4.6 shows possible interferometric techniques that have been used for interferometric measurement of phase conjugate fidelity. Figure 4.6(a) shows an interferogram between the incident laser and Stokes waves. Ideally, the interferogram should be taken in the same 'conjugate'

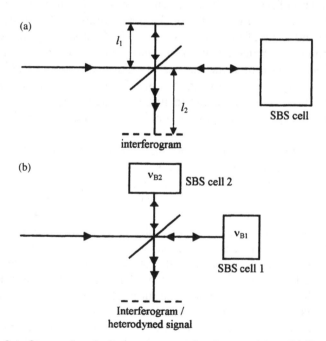

Figure 4.6. Interferometric techniques to measure the phase conjugate fidelity factor H (a) using an interferogram between laser and Stokes wavefronts and (b) by heterodyning the signals from two nearly identical SBS cells.

plane at the beamsplitter itself in Figure 4.6(a) $(l_1 + l_2 \approx 0)$, but if the incident laser has a nearly plane wavefront, its wavefront at the interferogram is similar to that at the beamsplitter. The interferogram will show the differences in the Stokes from a plane wave. Figure 4.6(b) shows a different approach in which the laser is equally split and sent to two SBS cells. If the SBS cells have a small difference in frequency shift the interferogram is a heterodyned signal that can be monitored on a fast photodetector, the difference frequency will be seen as a modulation on the combined signals. If the signals are perfectly spatially correlate the modulation will be maximal. If they are uncorrelated the modulation will tend to zero. The depth of modulation on the signal performs (with suitable normalization) the spatial correlation integral H, incorporating amplitude and phase.

4.4 Polarization properties of SBS phase conjugation

The polarization of the laser radiation has been assumed to be a pure linearly polarized state. The SBS process involves formation of an acoustic wave that acts as a diffractive mirror to backscatter the radiation into a phase conjugate under the correct selection criteria. In an isotropic medium the acoustic wave is a longitudinal wave and the density fluctuation is a scalar quantity such that no coupling can occur between fields of orthogonal polarization state.

Consider a laser and Stokes field with inclusion of polarization state

$$\mathbf{E}_L(\mathbf{r}) = E_L(\mathbf{r}) \cdot \mathbf{e}_S, \qquad \mathbf{E}_S(\mathbf{r}) = E_S(\mathbf{r}) \cdot \mathbf{e}_S \tag{4.13}$$

where $E_S(\mathbf{r})$ is the spatial form of the field and $\mathbf{e}_S$ is its polarization state. The acoustic field $Q(r)$ is scalar and has no polarization component. The interaction equations for the Stokes field and the acoustic amplitude are

$$\frac{\partial}{\partial z}\mathbf{E}_S - \frac{i}{2k}\nabla_T^2\mathbf{E}_S = \mathbf{E}_L Q^* \tag{4.14}$$

$$\left(\frac{\partial}{\partial t} + \frac{1}{2\tau_B}\right)Q = \gamma \mathbf{E}_L \cdot \mathbf{E}_S^*. \tag{4.15}$$

In equation (4.15), the laser and Stokes waves occur as a vector dot product $\mathbf{E}_L \cdot \mathbf{E}_S^* = E_L E_S(\mathbf{e}_L \cdot \mathbf{e}_S^*)$. If the laser and Stokes fields have orthogonal polarization $(\mathbf{e}_L \cdot \mathbf{e}_S^*) = 0$ and no acoustic field is generated; when they are parallel in polarization $(\mathbf{e}_L.\mathbf{e}_S^*) = 1$, and they form the maximum strength of acoustic field. In equation (4.14), the growth of the Stokes field is due to the scattering of the laser field $\mathbf{E}_L$ from the scalar acoustic field Q, and for the case of plane wave interaction the Stokes field will develop the same polarization state as the laser field $\mathbf{e}_S = \mathbf{e}_L$.

Hence for some special cases we expect linear laser polarization to generate linear Stokes polarization

$$\mathbf{e}_L = \mathbf{e}_x, \qquad \mathbf{e}_S = \mathbf{e}_x.$$

Right-hand circularly polarized laser induces left-hand circularly polarized Stokes that is parallel to the laser field polarization. (The change in handedness is purely a convention due to the change in direction of the Stokes field and a convention that the handedness is related to the rotation direction as perceived along the direction of propagation.)

$$\mathbf{e}_L = \frac{1}{\sqrt{2}}(\mathbf{e}_x + i\mathbf{e}_y), \qquad \mathbf{e}_S = \frac{1}{\sqrt{2}}(\mathbf{e}_x + i\mathbf{e}_y).$$

Note that in this case $(\mathbf{e}_L \cdot \mathbf{e}_S^*) = 1$, as expected for parallel polarization states.

We can extend the definition of phase conjugation to specify a process called vector phase conjugation (VPC) in which both spatial wavefront *and* polarization state are simultaneously reversed. The property of VPC is most directly stated by its ability for phase aberration correction—'if a laser field with a particular spatial distribution and polarization state passes through (reciprocal) phase components inducing changes to the spatial distribution and polarization state, the vector phase conjugate field is the one that on passage back through the same phase components will return the field to the original spatial distribution and polarization state'.

It is seen that for a pure polarization state, SBS induces a parallel Stokes polarization $\mathbf{E}_L(\mathbf{r}) = E_L(\mathbf{r}) \cdot \mathbf{e}_L$; $\mathbf{E}_S(\mathbf{r}) = r_c E_L^*(\mathbf{r}) \cdot \mathbf{e}_L$. With regard to polarization, SBS is therefore identical to the reflection property of a conventional mirror. The consequence is that the SBS process has the ability to phase conjugate the spatial distribution of the light but not the polarization state. SBS is not a vector phase conjugator. This can be useful for optical isolation of the laser from the SBS reflection whose phase conjugate form can seriously couple with the laser, even causing damage to optical components. Two optical isolation systems are commonly used: a polarizer and quarter-wave plate combination or a polarizer and Faraday rotator combination as shown in figure 4.7.

Figure 4.7(a) shows the simplest case of linear polarization (and more generally a polarizer with parallel transmission axis). The SBS cell maintains the linear polarization in reflection and the phase conjugate wave reproduces the original polarization state. In view of the definition of vector phase conjugation this is achieved, but since there is no polarization disturbance by the propagation path this is a trivial result. A counter example is as follows: circular polarized light passing through a quarter-wave plate to produce linear polarized light would reflect as linear polarized light and on second passage through the wave plate would become circular polarized light, but orthogonal to the original polarization, i.e. not the vector phase

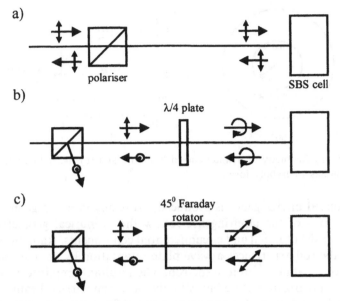

Figure 4.7. SBS polarization reflection properties. (a) Linear polarization case, (b) optical isolation produced polarizer and (reciprocal) quarter-wave retardation plate and (c) optical isolation using a (non-reciprocal) Faraday rotator.

conjugate, despite the reproduction of the linear polarization at the SBS cell.

Figure 4.7(b) shows the double pass of the quarter-wave retardation plate converting linear polarized light to orthogonal linear polarization that is rejected by the polarizer. This decouples the SBS reflection from the laser source. In figure 4.7(c), an equivalent isolation with a Faraday rotator is demonstrated. The Faraday rotator induces rotation of the polarization state of light by 45°, by use of a crystal in the presence of a strong magnetic field (the Faraday effect). Unlike a retardation plate, the Faraday rotator is an example of a non-reciprocal element; its rotation is independent of the direction of propagation through the device.

4.5 Thermally-induced lensing and depolarization in laser amplifiers

From the earliest discovery of SBS phase conjugation, an important application has been its use to compensate the phase distortions incurred by a laser beam on passage through an active laser amplifier. This distortion is produced normally by the intense pumping mechanism producing the inversion in the amplifying medium. In solid-state laser amplifiers, as well as phase distortion produced by thermally-induced refractive changes, a significant

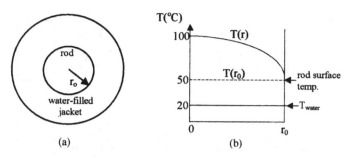

Figure 4.8. (a) Geometry of a water-cooled laser rod and (b) the induced temperature distribution has a parabolic form.

stress-induced birefringence is produced, especially in a rod geometry. The principal axes of the birefringence follow the cylindrical geometry of the rod and consist of radial and circumferential components. Radiation passing through the rod experiences a wave plate retardation by an amount dependent on the spatial position in the rod. The resultant amplified radiation is not a pure polarization state but is known as a depolarized beam.

Consider a laser rod, as shown in figure 4.8, such as in a lamp-pumped Nd:YAG laser. The temperature increase induced in the rod is given by the heat diffusion equation (in cylindrical coordinates)

$$\frac{d^2 T}{dr^2} + \frac{1}{r}\frac{dT}{dr} + \frac{Q}{\kappa} = 0 \quad \text{with } T(r_0) \text{ fixed} \tag{4.16}$$

where κ is the thermal conductivity and Q the input heat power per unit volume and assuming uniform heating $Q = P_h/(\pi r_0^2 l)$, is given by total heating power P_h divided by rod volume (r_0 and l are radius and length of rod, respectively). For a Nd:YAG laser typically $P_h \approx 5\%$ of electrical input (P_E) supplied to the lamp, producing inversion.

The heat diffusion equation has the solution

$$T(r) = T(r_0) + \left(\frac{Q}{4\kappa}\right)(r_0^2 - r^2) \tag{4.17}$$

which has a parabolic distribution. Due to the temperature-dependence of refractive index [e.g. $(dn/dT)_{YAG} = 7.3 \times 10^{-6} \, \text{K}^{-1}$] a parabolic variation in refractive index is induced, $\Delta n(r) = (dn/dT)\,\Delta T(r) = -(Q/4\kappa) \cdot (dn/dT)r^2$. This is equivalent to a lens which over the length of the rod has a focal power given by

$$\frac{1}{f} = \frac{Q \cdot l(dn/dT)}{2\kappa} = DP_h. \tag{4.18}$$

By way of example with Nd:YAG, with $(dn/dT)_{YAG} = 7.3 \times 10^{-6} \, \text{K}^{-1}$, $\kappa = 14 \, \text{W m}^{-1} \, \text{K}^{-1}$, $r_0 = 2 \, \text{mm}$, if $P_h = 5\% \, P_E$ (electrical input power) we deduce $1/f = 10^{-3} P_E(W)$ dioptres, i.e. $f = 1 \, \text{m}$ for $P_E = 1 \, \text{kW}$, and with

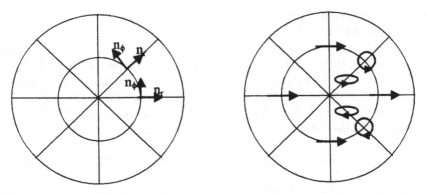

Figure 4.9. The stress-induced birefringence in a laser rod has (a) cylindrical principal axes of refractive index and for a linearly polarized input beam (b) leads to an output beam with spatially-dependent polarization known as a depolarized state.

laser efficiency $\eta = 1\%$ this occurs at output power $P_0 \sim 10\,\text{W}$. Hence, thermally-induced lensing needs to be considered in cavity design at a few watts and becomes very severe above a few tens of watts.

In addition to thermally-induced lensing, differential thermal expansion leads to parabolic stress σ in the rod, with radial (σ_r) and tangential (σ_ϕ) components given by $\sigma_r(r) = Q \cdot S(r^2 - r_0^2)$ and $\sigma_\phi(r) = Q \cdot S(3r^2 - r_0^2)$.

At sufficiently high power, the stress can fracture the rod but at lower powers the strain via the photoelastic effect induces a change in refractive index n. The stress-induced birefringence has cylindrical principal axes (see figure 4.9(a)). For Nd:YAG, with $n_0 = 1.82$, expansion coefficient $\alpha_0 = 7.5 \times 10^{-6}\,\text{K}^{-1}$, and $\kappa = 0.14\,\text{W\,cm}^{-2}\,\text{K}^{-1}$ this leads to $\Delta n_r = (-2.8 \times 10^{-6})Qr^2$; $\Delta n_\phi = (+0.4 \times 10^{-6})Qr^2$, where Q is in W/cm^2 and r is in cm. The birefringent rod acts as an effective retardation wave plate that increases as r^2 increases. Along 45° axes, maximum 'wave plate' retardation occurs. Linear polarized light entering the rod emerges depolarized as indicated in figure 4.9(b).

4.6 Vector phase conjugation of depolarized radiation via SBS

For a pure polarization state we can write $\mathbf{E}(\mathbf{r}) = E(\mathbf{r})\mathbf{e}$, whereas for a depolarized beam we must write $\mathbf{E}(\mathbf{r}) = E_1(\mathbf{r})\mathbf{e}_1 + E_2(\mathbf{r})\mathbf{e}_2$, where $\mathbf{e}_1$ and $\mathbf{e}_2$ are any orthogonal pair of polarization states ($\mathbf{e}_1 \cdot \mathbf{e}_2^* = 0$). It is most instructive to take the particular orthogonal polarization states that correspond to laser fields $E_1(\mathbf{r})$ and $E_2(\mathbf{r})$ that are uncorrelated when averaged over their transverse profile, i.e. $\langle E_1(\mathbf{r}) \cdot E_2^*(\mathbf{r}) \rangle = 0$. Introducing a parameter called the degree of polarization p, with values $0 < p < 1$, we have fields related to total intensity I by $\langle E \cdot E^* \rangle = I$; $\langle E_1 \cdot E_2^* \rangle = 0$; $\langle |E_1|^2 \rangle = I(1+p)/2$; $\langle |E_2|^2 \rangle = I(1-p)/2$. When one of the field components is zero, we have a pure polarization state ($p = 1$) and when they are equal we have a completely depolarized state ($p = 0$).

In this section we make use of equations (4.14) and (4.15) for the Stokes and acoustic field amplitudes, and take the Stokes field $\mathbf{E}_S(\mathbf{r}) = S_1(\mathbf{r})\mathbf{e}_1 + S_2(\mathbf{r})\mathbf{e}_2$ as an expansion in the same pair of orthogonal polarization states as the laser. In the steady state, the acoustic amplitude is given by $Q^* = 2\tau_B\gamma \cdot (E_1^*S_1 + E_2^*S_2)$ which when inserted in equation (4.14) leads to two equations for the Stokes polarization components

$$\frac{\partial}{\partial z}S_1 - \frac{i}{2k}\nabla_T^2 S_1 = -\frac{g_B}{2}[|E_1|^2 S_1 + E_1 E_2^* S_2] \qquad (4.19a)$$

$$\frac{\partial}{\partial z}S_2 - \frac{i}{2k}\nabla_T^2 S_2 = -\frac{g_B}{2}[|E_2|^2 S_2 + E_2 E_1^* S_1]. \qquad (4.19b)$$

The first gain term in equation (4.19a) is the normal SBS growth term for, as described before, linearly polarized radiation and leads to a two-fold higher exponential growth rate for a Stokes field distribution that is phase conjugate to the incident field ($S_1 \approx E_1^*$) compared with one that is uncorrelated. The second term in the right-hand side of equation (4.19a) is due to the scatter of one laser field E_1 from the acoustic wave component generated by the orthogonal field components E_2, and has an enhanced growth rate for a Stokes field that is correlated to the orthogonal field component ($S_1 \approx E_2^*$). Identical terms exist in equation (4.19b). The overall effect is that the total Stokes field is not phase conjugate to the incident laser field, let alone vector phase conjugate. In particular, neglecting the second terms, the first terms of equations (4.19) show the dominant solution as $S_1 \approx E_1^* \exp(\frac{1}{2}g_B(1 + p)l)$ and $S_2 \approx E_2^* \exp(\frac{1}{2}g_B(1 - p)l)$. For a weakly depolarized beam ($p \approx 1$) and $E_1 \gg E_2$, the Stokes output is predominantly the phase conjugate of the strongest laser polarization component and the other polarization is negligible, i.e. $S \approx E_1^* \exp(g_B/2(1 + p)l)$.

The above shows that SBS cannot vector phase conjugate depolarized radiation and furthermore the ordinary phase conjugation ability of SBS is also degraded, with the Stokes tending only to the conjugate of the strongest polarization component for weakly depolarized beams and for strongly depolarized beams having components due to scatter of one polarization from the acoustic wave generated by the orthogonal component. A depolarization compensation scheme has been devised, however, using the SBS process [4.36] as shown in figure 4.10. A polarizer is used to split depolarized laser radiation ($\mathbf{E}(\mathbf{r}) = E_1(\mathbf{r})\mathbf{e}_x + E_2(\mathbf{r})\mathbf{e}_y$) into two orthogonal linear polarization components, E_1 and E_2. A half-wave retardation plate rotates the polarization in one of the arms by 90°. The two beams E_1 and E_2 are incident on a common SBS cell, which now sees them as a pure polarization state and is able to form their phase conjugate. It can be shown that if the two beams are well overlapped in an SBS cell this leads to each of them experiencing equal reflectivity (even if one component is weaker than the other) and a

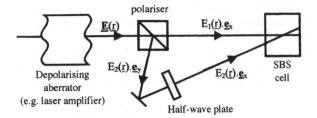

Figure 4.10. A configuration to produce vector phase conjugation of a depolarized laser beam (after ref [4.36]).

common phase shift. The Stokes component in arm two is rotated by 90° back to its original polarization and is recombined at the polarizer with the correct phasing. The resultant Stokes beam is the vector phase conjugate of the incident beam and will restore the radiation to its original spatial and polarization state on re-traversing the aberrator.

References

4.1 Zel'dovich B Ya, Pilipetsky N F and Shkunov V V 1985 *Principles of Phase Conjugation* (Berlin: Springer)
4.2 Zel'dovich B Ya, Popovichev V I, Ragaal'skii V V and Faizullov F S 1972 *Sov. Phys. JETP* **15** 109
4.3 Sidorovich V G 1976 *Sov. Phys. Tech. Phys.* **21** 1270
4.4 Yariv A 1978 *IEEE J. Quantum Electron.* **QE-14** 650
4.5 Levenson M D 1980 *Opt. Lett.* **5** 182; 1983 *J. Appl. Phys.* **54** 4305
4.6 Levenson M D, Johnson K M, Hanchett V C and Chiang K 1981 *J. Opt. Soc. Am.* **71** 737
4.7 Levenson M D and Chiang K 1982 *IBM J. Res. Dev.* **26** 160
4.8 Gower M C 1984 *Prog. Quantum Electron.* **9** 101; 1988 *J. Mod Optics* **35** 449
4.9 Gower M C 1994 High resolution image projection with phase conjugation mirrors in *Optical Phase Conjugation* ed M Gower and D Proch (Berlin: Springer) p 315
4.10 Pfau A K, Proch D and Bachmann F 1990 *Opt. Lett.* **15** 6
4.11 Sokolovaskaia A I, Brekhovskikh G I and Kudriavtseva A D 1983 *J. Opt. Soc. Am.* **73** 554; 1987 *IEEE J. Quantum Electron.* **QE-23** 1332
4.12 Lind R C, Steel D G and Dunning G J 1982 *Opt. Eng.* **21** 190
4.13 Kaiser W and Maier M 1972 *Laser Handbook* vol 2 ed F T Arrechi and E O Schultz (Amsterdam: North-Holland) p 1077
4.14 Basov N G, Zubarev I G, Kotov A V, Mikhailov S I and Smirnov M 1980 *Sov. J. Quantum Electron.* **9** 237
4.15 Andreev N F, Bespalov V I, Kiselev A M, Matreev A Z, Pasmanik G A and Shilov A A 1980 *JETP Lett.* **32** 625
4.16 Gross R W R, Amimoto S T and Garman-DuVall L 1989 *Opt. Lett.* **16** 94
4.17 Ippen E P and Stolen R H 1972 *Appl. Phys. Lett.* **21** 539
4.18 Agrawal G P 1995 *Nonlinear Fiber Optics* 2nd edition (San Diego: Academic Press) p 370

4.19 Zeldovich B Y 1998 *Overview of optical phase conjugation* Technical Digest, Summaries, Conference on Lasers and Electro-Optics Conference Edition **6** 559 (Washington: Optical Society of America)

4.20 Eichler H J, Haase A, Liu B and Mehl O 1998 *Laser-Physics* **8** 769

4.21 Kuzin E A, Petrov M P and Fotiadi A A 1994 Phase Conjugation by SMBS in *Optical Fibers in Optical Phase Conjugation* ed M Gower and D Proch (Berlin: Springer) p 74

4.22 Eichler H, Kunde J and Liu B 1997 *Opt. Commun.* **139** 327

4.23 Eichler H J, Liu B, Wittler O and Zhu Q 1998 *Saturation and oscillation of SBS reflectivity in fiber phase conjugators* Nonlinear Optics '98. Materials, Fundamentals and Applications Topical Meeting, New York p 468

4.24 Peral E and Yariv A 1999 *IEEE Quantum Electron.* **35** 1185

4.25 Eichler H J, Liu B, Haase A, Mehl O and Dehn A 1998 *Proc. SPIE* **3263** 20

4.26 Eichler H J, Haase A and Mehl O 1998 *Proc. SPIE* **3264** 9

4.27 Jun Chen, Zhi Hong, Chenfang Bao, Wenfa Qiu and Xiuping Wang 1998 *Proc. SPIE* **3549** 86

4.28 Dehn A, Eichler H J, Haase A, Mehl O and Schwartz J 1998 *Proc. SPIE* **3403** 65

4.29 Eichler H J, Haase A, Kunde J, Liu B and Mehl O 1997 *Power scaling of solid state lasers over 100 W with fiber phase conjugators* Solid State Lasers: Materials and Applications Sino-American Topical Meeting 168, Technical Digest, (Washington: Optical Society of America)

4.30 Eichler H J, Haase A and Menzel R 1996 *Opt. Quantum Electron.* **28** 261

4.31 Heuer and Menzel R 1998 *Opt. Lett.* **23** 834

4.32 Heuer and Menzel R 1999 *Stimulated Brillouin scattering in an internally tapered fiber* Proc. Int. Conf. LASERS '98, Soc. Opt. & Quantum Electron., McLean, VA p 1215

4.33 Harrison R G, Kovalev V I, Weiping Lu and Dejin Yu 1999 *Opt. Commun.* **163** 208

4.34 Eichler H J, Dehn A, Haase A, Liu B, Mehl O and Rücknagel S 1999 *Proc. SPIE* **3267** 158

4.35 Hellwarth R W 1978 *J. Opt. Soc. Am.* **68** 1050

4.36 Basov N G *et al* 1978 *JETP Lett.* **28** 197

Chapter 5

Solutions of the three-dimensional SBS model

A three-dimensional wave model for the stimulated Brillouin scattering can be built and analytically treated, in the case of slowly-varying-envelope approximation, in order to emphasize the transverse effects in SBS. The integration of the general SBS equation system can lead to more accurate analytical expressions for the pump and Stokes wave intensities and for SBS reflectivity. SBS can be modelled by a set of three differential equations describing the interaction between the light waves and the acoustical wave yielded in a nonlinear medium [5.1–5.10]. Due to the complexity of the non-linear differential equations of the three-dimensional SBS model, several authors have solved them in different approximations or numerically, giving information about the evolution in space and time of the three inter-acting waves: pump wave, Stokes wave and acoustic wave.

Ridley *et al* [5.11] studied the three-dimensional SBS amplification process when the pump intensity profile is Gaussian; they have obtained an analytical solution in the case of quasi-steady state and non-depleted pump, and numerical results in the case of depleted pump. Suni and Falk [5.12] and Miller *et al* [5.13] have done numerical simulations of two-dimensional and three-dimensional SBS, in steady state and non-depleted pump approximations, valid near the SBS threshold only. Numerical models for one-dimensional SBS, in depleted steady-state, were presented by Tang [5.14] and by Menzel and Eichler [5.15]. Three-dimensional SBS, in depleted steady state, was numerically studied by Kummrow [5.16] and Moore *et al* [5.17], who developed the light waves in terms of Hermite–Gauss orthonormal functions. Stoddard *et al* [5.18] developed an analytic model for the evaluation of the scattering cross-section, which depends on the aperture, natural and induced non-uniformities. In the study of the transverse effects in SBS, Visnyauskas *et al* [5.19] constructed a simplified three-dimensional model (with cylindrical symmetry) and found an analytical solution for an initial condition taken as a series of Gauss–Laguerre polynomials. The solution permits the correlation of some transversal and longitudinal effects (for example, the dependence of Stokes pulse duration on the divergence angle and the dependence of SBS reflectivity on the transversal envelope of the pump intensity). Raab *et al* [5.20] studied the transversal effects in optical phase conjugation in lasers with an SBS mirror. The expansion in Gauss–Hermite modes is performed and the transversal problem is linearized for the clarity of the solutions. Babin *et al* [5.21] and Vlad *et al* [5.22] have built a three-dimensional wave model for SBS and they have analytically treated the SBS process, in the case of slowly-varying-envelope approximation and steady-state regime. Kuzin *et al* [5.23] and Mashkov and Temkin [5.24] studied the transversal eigenmodes propagation in different SBS waveguiding structures. Rae *et al* [5.25] used a numerical model for SBS in optical fibres with non-uniform properties, which in turn induce spatial non-uniformities of the induced acoustic field. Anikeev *et al* [5.26], Lehmberg [5.27] and Hu *et al* [5.28], studied numerically three-dimensional steady-state SBS, with depleted pump, in optical fibres, where electric fields are expanded in the series of orthonormal functions corresponding to the propagation modes of the fibres.

Recently, Afshaarvahid *et al* [5.29] have presented a transient three-dimensional model of SBS and have used it to study the phase conjugation in SBS and the mode structure of the Stokes and pump pulse inside the SBS cell. They confirm the experimental observation of pulse-shape dependence of SBS phase conjugation fidelity presented by Dane *et al* [5.30].

In this chapter, we find the analytical solution of SBS with a depleted spatial Gaussian pump beam, in steady-state, which can lead to high reflectivity and fidelity. Merit factors are calculated and the analytical results compared with the experimental and numerical ones. More generally, we present analytical solutions of the SBS three-dimensional model in the case of pump beams with axial symmetry [5.21, 5.22]. Finally, the transient three-dimensional numerical simulations of Afshaarvahid and Munch [5.29] are shown in order to provide a general model for SBS and some effects of transient phenomena related to the reflectivity and fidelity.

5.1   SBS model with a spatial Gaussian pump beam

The three-dimensional modelling of SBS can start with the case of the pump wave with Gaussian space profile:

$$E_L(r, z) = E_{L0}(z) \exp\left(-\frac{r^2}{w_L^2(z)}\right) \tag{5.1a}$$

and with the intensity

$$|E_L(r, z)|^2 = I_L(r, z) = I_L(0, z) \exp\left(-\frac{2r^2}{w_L^2(z)}\right) \tag{5.1b}$$

where w_L is the beam radius at the $1/e$ point of the field relative to axial value. We assume that the scattered Stokes beam may have a similar spatial distribution and look for a solution of the form

$$E_S(r,z) = E_{S0}(z) \exp\left[-\frac{i}{2} \cdot Q(z) \cdot r^2\right], \qquad Q(z) = \frac{K_S}{R_S(z)} - \frac{2i}{w_S^2(z)} \qquad (5.2)$$

where Q is a complex curvature, $R_S(z)$ is the radius of curvature of the wavefront, $w_S(z)$ is the Gaussian beam width (spot size) and K_S is the wavevector. This solution is similar to that of the free-space propagation equation, except the dependencies of E_{S0} and Q on z. Further, considering that the diffraction of the Stokes wave is weak (the low divergence approximation), one can take $(1/R_S) \rightarrow 0$ and write

$$E_S(r,z) \approx E_{S0}(z) \exp\left(-\frac{r^2}{w_S^2(z)}\right). \qquad (5.3)$$

We shall use the SBS equations (1.28) and (1.29) in the steady-state approximation

$$\frac{\partial E_L}{\partial z} + i\frac{1}{2K_L}\nabla_\perp^2 E_L + \frac{1}{2}\alpha E_L = -\frac{1}{2}g_B I_S E_L$$

$$\frac{\partial E_S}{\partial z} - i\frac{1}{2K_S}\nabla_\perp^2 E_S - \frac{1}{2}\alpha E_S = -\frac{1}{2}g_B I_P E_S \qquad (5.4)$$

where K_L and K_S are the optical wavevectors of pump and Stokes fields and can be taken to be equal. The steady-state regime occurs when the pump duration t_L (or more generally when the time variation of the pump) is longer than the acoustic response time $t_L > \tau_B \cdot (g_B I_L L)^{1/2}$ and the time derivatives in the SBS equations can be neglected. For simplicity, absorption shall be neglected ($\alpha \approx 0$), and the normalized complex amplitudes of the pump and Stokes waves shall be introduced:

$$E_L' = \sqrt{\frac{cn\varepsilon_0}{2I_0}}E_L, \qquad E_S' = \sqrt{\frac{cn\varepsilon_0}{2I_0}}E_S \qquad (5.5)$$

with $I_0 = I_L(0,0)$ is the maximum pump intensity on the axis and at the entrance to the SBS medium. Passing to the cylindrical coordinates transforms the SBS equations to the following system

$$\frac{\partial E_L'(z,r)}{\partial z} + \frac{i}{2K}\left(\frac{\partial^2 E_L'(z,r)}{\partial r^2} + \frac{1}{r}\frac{\partial E_L'(z,r)}{\partial r}\right) = -(\sigma_B)|E_S'(z,r)|^2 E_L'(z,r)$$

$$\frac{\partial E_S'(z,r)}{\partial z} - \frac{i}{2K}\left(\frac{\partial^2 E_S'(z,r)}{\partial r^2} + \frac{1}{r}\frac{\partial E_S'(z,r)}{\partial r}\right) = -(\sigma_B)|E_L'(z,r)|^2 E_S'(z,r) \qquad (5.6)$$

with $\sigma_B = g_B I_0/2$. We shall denote

$$I_L = |E'_L|^2, \qquad I_S = |E'_S|^2. \tag{5.7}$$

The solutions of the SBS equation system (5.6) can be found for $I_L(z)$, $I_S(z)$, $w_L(z)$ and $w_S(z)$ [5.32] and using these results, we can calculate an overall parameter, the SBS (energy) reflectivity, which is simple to measure experimentally:

$$R = \left(\frac{\varepsilon_S}{\varepsilon_L}\right) \tag{5.8}$$

where ε_L and ε_S are the energies of the pump and Stokes pulse, respectively.

In the case of weak diffraction, the intensity of the pump field, with Gaussian spatial and temporal profiles, can be written as

$$I_L(r,z,t) \approx I_0 \exp\left(-2\frac{r^2}{w_L^2}\right) \exp\left(-\frac{t^2}{t_L^2}\right) \tag{5.9}$$

where $I_0 = I_L(0)$ is the peak intensity (at $r = 0$, $z = 0$), $w_L(z)$ is the pump beam radius and t_L is the laser pulse duration.

The intensity of the Stokes beam may be considered also as product of spatial and temporal factors, when the strength of the nonlinear coupling is small:

$$I_S(r,z,t) \approx I_{S0}(z) \exp\left(-2\frac{r^2}{w_S^2(z)}\right) \exp\left(-\frac{t^2}{t_S^2}\right)$$

$$= f_1[r, w_L, (w_S/w_L)\ldots] \cdot f_2[t, t_L, (t_S/t_L), \ldots] \cdot I_L \tag{5.10}$$

where $f_1[r, w_L, (w_S/w_L)\ldots]$ is a function related to the phase-conjugation fidelity $(f_1 \rightarrow 1, w_S(z)/w_L \rightarrow 1)$, and the function $f_2[t, t_L, (t_S/t_L), \ldots]$ is related to the laser pulse compression $(f_2 \rightarrow 1, t_S/t_L \rightarrow 1)$.

The pump and Stokes energies and the SBS reflectivity may be written as

$$\varepsilon_L = 2\pi I_0 \int_0^\infty \exp\left(-\frac{2r^2}{w_L^2}\right) \rho\,d\rho \cdot \int_0^\infty \exp\left(-\frac{t^2}{t_L^2}\right) dt \propto I_0 w_L^2 t_L \tag{5.11}$$

$$\varepsilon_S \propto I_{S0} w_S^2 t_S \tag{5.12}$$

$$R = \frac{\varepsilon_S}{\varepsilon_L} = \frac{I_{S0}(z) w_S^2(z)}{I_0 w_L^2} \frac{t_S}{t_L}. \tag{5.13}$$

In equation (5.13) for energy reflectivity, we can identify the first factor to a transversal component of the SBS reflectivity and the second factor to pulse compression. In the steady state, where the pulse compression is always close to unity, the dependence of R on the pump energy is almost saturated and essentially determined by the transverse features of the Stokes wave.

In the case of negligible pump pulse compression, which is expected for long pump pulses, the SBS reflectivity can be evaluated by

$$R = \left| \frac{I_S(z)w_S^2(z)}{I_L(z)w_L^2(z)} \right|. \tag{5.14}$$

After the calculation of the integration constants, in the small SBS reflectivity (the pump non-depletion) regime, one can introduce approximations such that the reflectivity (at $z = L$) takes the simple form [5.32]

$$R \approx g_B I_0 L \frac{I_S(0) - I_S(L)}{w_S^{-2}(0) - w_S^{-2}(L)}, \tag{5.15}$$

which shows a linear dependence on the pump intensity ($I_0 \propto \varepsilon_L$). We remark that, for pump short pulse duration (shorter than the phonon life-time), the pulse compression could occur, which multiplies the reflectivity from equation (5.14) by a factor $1/\sqrt{I_0}$ and leads to a parabolic dependence, $R \propto \sqrt{I_0}$.

In the steady-state saturation regime (depleted pump case), the reflectivity can be expressed as

$$R \approx \frac{w_S^2(L)}{w_L^2(L)} \left[1 - \frac{1}{2(g_B I_0 L)^2} \frac{w_L^4(0)}{w_L^4(L)} \right]. \tag{5.16}$$

The Stokes beam is narrowed due to the higher gain at the centre of the pump beam, which ensures the inequality $R < 1$ for equation (5.16).

The analytical results can be checked with experimental results. Figure 5.1 shows a Nd:YAG laser (oscillator–amplifier system), the measuring and coupling systems and an SBS cell. The oscillator consists of a Nd:YAG laser, Q-switched operated using a Pockels cell with KDP crystal. The output was in a near diffraction limited beam quality with 5 mm diameter and in a pulse with a time duration of 60 ns. The pulse energy was increased by Nd:YAG amplifier and was focused with convergent lenses onto a cell containing CS_2 as the nonlinear medium. An optical isolator (a Glan prism polarizer and a quarter-wave plate) was introduced

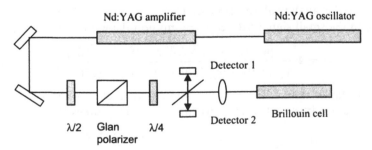

Figure 5.1. Experimental setup for the measurement of SBS reflectivity.

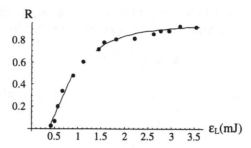

Figure 5.2. The dependence of the SBS reflectivity, R, on the Gaussian pump energy (including pump depletion). Experimental data are marked by points and, in the small signal region, the linear dependence derived in equation (5.15) is drawn with a continuous line, and in the saturation region the analytical dependence from equation (5.16) is used.

to decouple the SBS reflection from returning to the amplifier. In the experiment, the focusing lens had a focal length of 100 mm. In order to measure the SBS reflectivity, a wedge or a mirror was placed into the beam to get a reference for the pump energy and the backscattered Stokes energy using power meters, and pulse durations were measured using a fast photodiode and an oscilloscope.

In figure 5.2, the dependence of the SBS reflectivity, R, as a function of the pump energy is shown. The reflectivity increases with the energy of the incident light pulse, saturating at approximately 90%. The SBS threshold energy is estimated at 0.4 mJ, by extrapolating the experimental curve of the SBS reflectivity. One can note that good agreement between the calculated reflectivity and the experimental data is obtained for both the small signal and the saturation regimes.

The fidelity of the phase conjugated Stokes wave can be derived from the definition given by Zeldovich *et al* [5.1]

$$H(z) = \frac{\left|\int\int_{\Sigma} E_L(z,r)E_S(z,r)\,dr^2\right|^2}{\int\int_{\Sigma}|E_L(z,r)|^2\,dr^2 \int\int_{\Sigma}|E_S(z,r)|^2\,dr^2}. \qquad (5.17)$$

With the analytical solution of (5.16), it is possible to evaluate the fidelity in the small signal SBS case (pump non-depletion), at $z = L$, as

$$H = 4\left|\frac{[w_L^{-2}(0) - w_L^{-2}(L)][w_S^{-2}(0) - w_S^{-2}(L)]}{\{[w_L^{-2}(0) - w_L^{-2}(L)] + [w_S^{-2}(0) - w_S^{-2}(L)]\}^2}\right| \qquad (5.18)$$

which is smaller than 1, for $w_S < w_L$ and could reach 1, when $w_S = w_L$.

In the steady-state (saturation) case, the fidelity takes the form

$$H = \frac{4w_L^2(L)w_S^2(L)}{[w_L^2(L) + w_S^2(L)]^2} = \frac{4w_S^2(L)/w_L^2(L)}{[1 + w_S^2(L)/w_L^2(L)]^2}. \qquad (5.19)$$

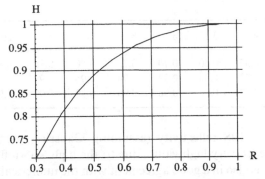

Figure 5.3. The dependence of the SBS fidelity, H, on the SBS reflectivity, R, for Gaussian pump waves (including pump depletion).

Taking into account the relation between the beam size ratio and the reflectivity from equation (5.14), we can further find the simple relation

$$H = \frac{4R}{(1 + R)^2}, \qquad R < 1 \tag{5.20}$$

which shows that fidelity grows to 1 faster than reflectivity and tends to 1, as $R \rightarrow 1$. This dependence is shown in figure 5.3.

These analytical results are in qualitative agreement with the simultaneous experimental results for SBS reflectivity R_{exp} and phase conjugation fidelity H_{exp}, obtained by Dane *et al* [4.30], with a single-frequency TEM$_{00}$ Q-switched Nd:YLF laser ($\lambda = 1053$ nm, $t_L = 15$ ns), in liquid carbon tetrachloride ($\tau_B \approx 1$ ns) and gaseous N$_2$ at 90 atm ($\tau_B \approx 15$ ns). In table 5.1, a comparison is made between experiment and theory with the fidelity, H_t, calculated with equation (5.20) using the values R_{exp}, and R_t is calculated with equation (5.14) for the saturation regime. The theoretical data compare favourably with the experimental ones, H_{exp}.

One remarks that, once the SBS reflectivity is measured, the fidelity could be estimated by equation (5.20), at least when the pump beam duration is much larger than the phonon lifetime (and the fidelity fluctuations are small). Thus, we could predict the fidelity in experiments devoted to the

Table 5.1.

ε_p (mJ)	10	20	30	40	50	60	70	80
R_{exp}	0	0.30	0.45	0.54	0.60	0.63	0.67	0.70
R_t	0	–	0.40	0.55	0.62	0.66	0.68	0.70
H_{exp}	0	0.73	0.85	0.90	0.93	0.94	0.96	0.97
H_t	0	0.71	0.86	0.91	0.94	0.95	0.96	0.97

Table 5.2.

$\varepsilon_L/\varepsilon_{L,th}$	2	3	6	8	10	15	20	30
R_{sim}	0.35	0.50	0.70	0.75	0.80	0.86	0.88	0.95
H_{sim}	0.77	0.85	0.92	0.94	0.95	0.97	0.97	0.98
H_t	0.77	0.84	0.96	0.98	0.99	0.99	0.99	0.99

measurement of SBS reflectivity (assuming Gaussian pump beam). The fidelity could reach very high values due to the high SBS reflectivity.

This result is also in a good agreement with the numerical calculations of Afshaarvahid and Munch [5.29]. For standard experimental conditions, with a pump pulse of 30 ns, $w_L = 0.5$ cm and energy 140 mJ and for an SBS material with the refractive index $n = 1$ and phonon lifetime 0.85 ns, the calculated overall reflectivity reaches 78% (at saturation) and the fidelity reaches 94%. Equation (5.20) and the above reflectivity lead to $H = 0.98$. Using the numerical calculated data from published data (figure 5.7 of [5.29]), denoted by R_{sim} and H_{sim} respectively, analytical relation between SBS reflectivity and fidelity from equation (5.20) can be checked. The predictions for the fidelity are denoted by H_t and are shown in table 5.2. One could remark that the agreement of prediction for fidelity (equation (5.24)) holds well with the numerical calculations even in the non-stationary SBS case, for Gaussian pump waves.

5.2   A three-dimensional model for non-stationary SBS and numerical results

The steady-state three-dimensional models may be successfully used when the transient phenomena can be neglected. However, there are some cases, as in high-power short-pulse lasers, in which the transient phenomena are important and require a more detailed description. These phenomena include threshold condition and reduction of fidelity in the case of using a pulse duration comparable with the phonon lifetime of the SBS material.

There are some reports on the transient regime of SBS for one-dimensional and three-dimensional models both in cell geometries and in waveguides. Afshaarvahid and Munch [5.29] have built a transient three-dimensional model for SBS, where transverse components of the electric fields are expanded in terms of Gauss–Hermite or Gauss–Laguerre functions (modes). This model assumes that the SBS process is initiated from a Gaussian random noise (as we have done in our one-dimensional SBS model), that the pump may be depleted and that transient phenomena are influencing the SBS fidelity. The corresponding SBS equations are numerically solved in order to calculate and present the dependences of the reflectivity and fidelity on the pump time evolution, energy and spatial structure.

We shall describe the model developed by Afshaarvahid and Munch and their principal numerical results. The authors start with the SBS equations for a non-stationary regime given by Kaiser and Maier [5.3] of the form

$$\left[\frac{i}{2K_L}\Delta_\perp + \left(\frac{n}{c}\frac{\partial}{\partial t} + \frac{\partial}{\partial z}\right)\right]E_L = ig_2 Q E_S$$

$$\left[\frac{i}{2K_S}\Delta_\perp + \left(\frac{n}{c}\frac{\partial}{\partial t} - \frac{\partial}{\partial z}\right)\right]E_S = -ig_2 Q^* E_L \qquad (5.21)$$

$$\left(\frac{\partial}{\partial t} + \Gamma_B\right)Q = ig_1 E_L E_S^*$$

where Q is the acoustic wave amplitude in suitably scaled units and g_1 and g_2 are coupling constants. In chapter 3, we have also identified the derivatives of the characteristics ξ_L, ξ_S as

$$\left(\frac{n}{c}\frac{\partial}{\partial t} + \frac{\partial}{\partial z}\right) = 2\frac{\partial}{\partial \xi_L}$$

$$\left(\frac{n}{c}\frac{\partial}{\partial t} - \frac{\partial}{\partial z}\right) = 2\frac{\partial}{\partial \xi_S} \qquad (5.22)$$

$$\xi_L = \frac{c}{n}t + z, \qquad \xi_S = \frac{c}{n}t - z.$$

The transversal components of the light (electric) waves are developed in series of orthonormal modes (bases of functions)

$$E_L(\mathbf{r}, z, t) = \sum_m a_m(z, t) A_m(\mathbf{r}, z)$$

$$E_S(\mathbf{r}, z, t) = \sum_m b_m(z, t) B_m(\mathbf{r}, z) \qquad (5.23)$$

where $a_m(z, t)$ and $b_m(z, t)$ are the coefficients of the series development and $A_m(\mathbf{r}, z)$ and $B_m(\mathbf{r}, z)$ form the spatial orthonormal mode bases. Introducing equations (5.23) into the SBS equations (5.21), one can obtain the equation set

$$\sum_m A_m \left(\frac{n}{c}\frac{\partial}{\partial t} + \frac{\partial}{\partial z}\right)a_m = ig_2 Q E_S$$

$$\sum_m B_m \left(\frac{n}{c}\frac{\partial}{\partial t} - \frac{\partial}{\partial z}\right)b_m = -ig_2 Q^* E_L \qquad (5.24)$$

$$\left(\frac{\partial}{\partial t} + \Gamma_B\right)Q = ig_1 \sum_{i,j} A_i(\mathbf{r}, z) B_j^*(\mathbf{r}, z)$$

$$\times [a_i(z, t) b_j^*(z, t) - f_{ij}(z, t)]$$

where $f_{ij}(z,t)$ represents a Gaussian random function, with zero average value, which characterizes the spontaneous Brillouin scattering (noise). The second-order correlation of this function is

$$\langle f_{ij}(z,t)f_{kl}^*(z',t')\rangle = Q_0\delta_{ik}\delta_{jl}\delta(z-z')\delta(t-t') \tag{5.25}$$

where $Q_0 = 2k_{\mathrm{B}}T\rho_0\Gamma_{\mathrm{B}}v_a^{-2}$, k_{B} is the Boltzmann constant, T is the temperature of the SBS material, ρ_0 is the average density, Γ_{B} is the Brillouin linewidth, v_a is the hypersound velocity in the SBS material and δ is the Dirac distribution. Equation (5.24) shows that the spontaneous Brillouin scattering could be represented as an ensemble of elementary independent scattering processes with zero average (a poly-stochastic process, which is specific to a homogenous Markov one). Equation (5.25) indicates the moment and the position in the SBS process, at which the spontaneous Brillouin scattering passes to the stimulated one.

The acoustic field in the last equation of (5.24) can be integrated as

$$Q(\mathbf{r},z,t) = ig_1\sum_{i,j} A_i(\mathbf{r},z)B_j^*(\mathbf{r},z)C_{ij}(z,t) \tag{5.26}$$

with

$$C_{ij}(z,t) = \int_0^t [a_i(z,\tau)b_j^*(z,\tau) + f_{ij}(z,\tau)]e^{-\Gamma_{\mathrm{B}}(t-\tau)}\,\mathrm{d}\tau.$$

Introducing this expression (5.26) into the first two equations of (5.24), one can obtain the well-known system of equations which characterize the evolution of the light interacting fields: the pump (E_{L}) and the Stokes (E_{S}), represented in the mode space

$$\sum_m A_m\left(\frac{n}{c}\frac{\partial}{\partial t} + \frac{\partial}{\partial z}\right)a_m = -g_1g_2\sum_{i,j,k} A_iB_j^*B_kC_{ij}b_k \tag{5.27}$$

and

$$\sum_m B_m\left(\frac{n}{c}\frac{\partial}{\partial t} - \frac{\partial}{\partial z}\right)b_m = -g_1g_2\sum_{i,j,k} A_i^*A_kB_jC_{ij}^*a_k. \tag{5.28}$$

Using the ortho-normalization property of the modes A_m and B_m, for example:

$$\int_{-\infty}^{+\infty} A_mA_n^*\,\mathrm{d}r^2 = \delta_{mn} \tag{5.29}$$

one can multiply both sides of equations (5.27) and (5.28) by A_m^* and B_m^*, respectively and integrate in the transversal plane in order to obtain the evolution equations of the interacting waves

$$\left(\frac{n}{c}\frac{\partial}{\partial t}+\frac{\partial}{\partial z}\right)a_m = -g_1g_2\sum_{i,j,k}C_{ij}b_k\int_{-\infty}^{+\infty}A_iB_kB_j^*A_m^*\,\mathrm{d}r^2 \tag{5.30}$$

$$\left(\frac{n}{c}\frac{\partial}{\partial t}-\frac{\partial}{\partial z}\right)b_m = -g_1g_2\sum_{i,j,k}C_{ij}^*a_k\int_{-\infty}^{+\infty}A_i^*A_kB_jB_m^*\,\mathrm{d}r^2. \tag{5.31}$$

Observing that, in equations (5.30) and (5.31) the integrals could be written as the tensors

$$g_{ijkm}^{\prime*}(z) = \int_{-\infty}^{+\infty}A_i(\mathbf{r},z)B_k(\mathbf{r},z)B_j^*(\mathbf{r},z)A_m^*(\mathbf{r},z)\,\mathrm{d}r^2$$

$$g_{ijkm}(z) = \int_{-\infty}^{+\infty}A_i^*(\mathbf{r},z)A_k(\mathbf{r},z)B_j(\mathbf{r},z)B_m^*(\mathbf{r},z)\,\mathrm{d}r^2 \tag{5.32}$$

we can further describe the SBS process by

$$\left(\frac{n}{c}\frac{\partial}{\partial t}+\frac{\partial}{\partial z}\right)a_m = -(g_1g_2)\sum_{i,j,k}\int_0^t\mathrm{d}\tau\,\mathrm{e}^{-\Gamma_B(t-\tau)}$$
$$\times\,[a_i(z,\tau)b_j^*(z,\tau)+f_{ij}(z,\tau)]b_k(z,\tau)g_{ijkm}^{\prime*}(z)$$

$$\left(\frac{n}{c}\frac{\partial}{\partial t}-\frac{\partial}{\partial z}\right)b_m = -(g_1g_2)\sum_{i,j,k}\int_0^t\mathrm{d}\tau\,\mathrm{e}^{-\Gamma_B(t-\tau)}$$
$$\times\,[a_i^*(z,\tau)b_j(z,\tau)+f_{ij}(z,\tau)]a_k(z,\tau)g_{ijkm}(z). \tag{5.33}$$

In the particular case of the steady-state regime ($\partial/\partial t = 0$), equations (5.33) take the form

$$\frac{\partial a_m}{\partial z} = -\left(\frac{g_1g_2}{\Gamma_B}\right)\sum_{i,j,k}[a_i(z)b_j^*(z)+f_{ij}(z)]b_k(z)g_{ijkm}^{\prime*}(z)$$

$$\frac{\partial b_m}{\partial z} = \left(\frac{g_1g_2}{\Gamma_B}\right)\sum_{i,j,k}[a_i^*(z)b_j(z)+f_{ij}(z)]a_k(z)g_{ijkm}(z). \tag{5.34}$$

The SBS transversal effects occur, in these equations, in the tensors $g_{ijkm}^{\prime*}(z)$ and $g_{ijkm}(z)$. These tensors indicate the manner in which the energy of the pump and the Stokes waves is distributed over their specific orthogonal modes.

SBS is one of the most important nonlinear optical processes which create the wavefront conjugation. Thus, we shall make the assumption that the modes of the Stokes wavefront are complex conjugated to those of the pump wavefront (for all indexes):

$$B_m(\mathbf{r},z) = A_m^*(\mathbf{r},z) \tag{5.35}$$

which will allow the separate evaluation of the energy scattered (reflected) from a pump mode into its conjugate (Stokes) one. From these evaluations, the SBS reflectivity and fidelity could be calculated.

The orthonormal modes may be represented by Gauss–Hermite functions in Cartesian coordinates [5.16, 5.17]

$$A_n(x, z) = \left(\frac{2}{\pi}\right)^{1/4} (2^n n! w(z))^{-1/2} \exp\left[i\left(n + \frac{1}{2}\right)\Psi(z)\right] H_n\left(\frac{\sqrt{2}x}{w(z)}\right)$$

$$\times \exp\left[-i\frac{Kx^2}{2R(z)} - \frac{x^2}{w^2(z)}\right] \tag{5.36}$$

and Gauss–Laguerre functions in cylindrical coordinates [5.13]

$$A_n(\mathbf{r}, z) = \left(\frac{2}{\pi}\right)^{1/2} \frac{1}{w(z)} \exp\left[i\left(n + \frac{1}{2}\right)\Psi(z)\right] L_n\left(\frac{2r^2}{w^2(z)}\right)$$

$$\times \exp\left[-i\frac{Kr^2}{2R(z)} - \frac{r^2}{w^2(z)}\right] \tag{5.37}$$

with

$$w^2(z) = w_0^2\left[1 + \left(\frac{z}{z_R}\right)^2\right], \qquad R(z) = z + \frac{z_R^2}{z},$$

$$\Psi(z) = \tan^{-1}\left(\frac{z}{z_R}\right) \quad \text{and} \quad r = \sqrt{x^2 + y^2}. \tag{5.38}$$

The Rayleigh distance, $z_R = \pi(n/\lambda)w_0^2$, is defined, for the fundamental mode, by the condition $w(z) = w_0$, where the beam waist, w_0, is related to the fundamental mode radius, w_i and the lens focal length, f, by: $w_0 = w_i(f/z_R)[1 + (f/z_R)^2]^{-1/2}$.

Assuming the cylindrical symmetry in the SBS geometry, the model may be built using Gauss–Laguerre functions [5.64]. In this case, the tensors from equations (5.32) become identical to the symmetric real gain tensor:

$$g_{ijkn} = \int_{-\infty}^{+\infty} A_i^* A_k B_j B_n^* \, \mathrm{d}r^2 = \frac{\exp[i(k - n - i - j)\Psi(z)]}{w^2(z)} \varepsilon_{ijkn} \tag{5.39}$$

where

$$\varepsilon_{ijkn} = \left(\frac{2}{\pi}\right) \int_0^{+\infty} e^{-2x} L_i(x) L_j(x) L_k(x) L_n(x) \, \mathrm{d}x.$$

Equation (5.39) shows that the gain tensor of any mode, g_{ijkn}, depends on a phase factor and on the mode coupling constant, ε_{ijkn}. The mode coupling constants, ε_{ijkn}, can be numerically calculated [5.29] and a recurrence relation could be found for them using the recurrence property of the Gauss–Laguerre function, which has the form

$$(n + 1)L_{n+1}(x) = (2n + 1 - x)L_n(x) + nL_{n-1}(x). \tag{5.40}$$

The recurrence relation for the coupling tensors, ε_{ijkn}, results as

$$\varepsilon_{ijk,n-1} = \frac{[3n+1-i-j-k]}{2(n+1)}\varepsilon_{ijkn} - \frac{n}{2(n+1)}\varepsilon_{ijk,n-1} + \frac{1}{2(n+1)}$$

$$\times \left[i\varepsilon_{i-1,jkn} + j\varepsilon_{i,j-1,kn} + k\varepsilon_{ij,k-1,n}\right].$$

The maximum value of the coupling tensors is obtained for the fundamental mode, $\varepsilon_{0000} = 0.3183$ and their other values are smaller the higher their indexes (for example, the tensor values for indexes higher than 1 are smaller than half the mentioned value). Moreover, the tensor values of the modes with equal indexes are double those corresponding to their adjacent modes.

Introducing equation (5.39) into equations (5.30) and (5.31), one can derive the general SBS model:

$$\left(\frac{n}{c}\frac{\partial}{\partial t} + \frac{\partial}{\partial z}\right)a_m = -g_1 g_2 \sum_{i,j,k} C_{ij} b_k g^*_{ijkm}(z) \qquad (5.42)$$

$$\left(\frac{n}{c}\frac{\partial}{\partial t} - \frac{\partial}{\partial z}\right)b_m = -g_1 g_2 \sum_{i,j,k} C^*_{ij} a_k g^*_{ijkm}(z). \qquad (5.43)$$

This equation system was numerically solved in [5.29] to determine the SBS reflectivity and fidelity.

Particularly, in the steady-state regime (when also f_{ij} is neglected), equations (5.42) and (5.43) become

$$\frac{\partial a_m}{\partial z} = -\left(\frac{g_1 g_2}{\Gamma_B}\right) \sum_{i,j,k} a_i(z) b^*_j(z) b_k(z) g'^*_{kmij}(z)$$

$$\frac{\partial b_m}{\partial z} = \left(\frac{g_1 g_2}{\Gamma_B}\right) \sum_{i,j,k} a^*_i(z) b_j(z) a_k(z) g_{kmij}(z). \qquad (5.44)$$

Moreover, if we consider in this regime a single-mode pump, a_s, the SBS equations take the form

$$\frac{\partial a_s}{\partial z} = -\left(\frac{g_1 g_2}{\Gamma_B}\right) \sum_{s,j,k} a_s b^*_j b_k g'^*_{kssj}$$

$$\frac{\partial b_m}{\partial z} = \left(\frac{g_1 g_2}{\Gamma_B}\right) \frac{|a_s|^2}{w^2(z)} \sum_j \varepsilon_{smsj} e^{-i(m+j)\Psi(z)} b_j(z). \qquad (5.45)$$

Above the threshold, it is usually assumed that the phase conjugated mode suppresses the other modes, i.e. $b_n \ll b_m$, for $n \neq m$. In this case, equation (5.45) leads to

$$\frac{\partial b_m}{\partial z} \approx \left(\frac{g_1 g_2}{\Gamma_B}\right) \frac{|a_s|^2}{w^2(z)} \varepsilon_{mmmm} b_m(z). \qquad (5.46)$$

Equation (5.46) admits a solution for the Stokes conjugated modes of the form

$$b_m(z) \propto \exp\left[\left(\frac{g_1 g_2}{\Gamma_B}\right) \frac{|a_s|^2}{w^2(z)} \varepsilon_{mmmm} z\right]. \qquad (5.47)$$

This solution shows that the longitudinal change of the Stokes modes depends on the pump intensity and the coupling constants. The smaller values of the coupling constants ε_{mmmm} lead to smaller reflectivity for the higher order modes, with the effect of their suppression in the competition with the conjugated single-mode, for the SBS gain.

For the Gauss–Laguerre fundamental mode, which is a perfect spatial Gaussian beam and for a Gaussian temporal pulse, the quality of the SBS phase conjugation may be evaluated by the fidelity factor of Zeldovich *et al*, which takes the form

$$H(z, t) = \frac{\left|\sum_i a_i(z, t) b_i(z, t)\right|^2}{\left(\sum_i |a_i(z, t)|^2\right)\left(\sum_i |b_i(z, t)|^2\right)} \qquad (5.48)$$

where $a_i(z, t)$ are the pump mode amplitudes and $b_i(z, t)$ are the Stokes mode amplitudes. This factor of fidelity is related to the pattern recognition by correlation and when $a_i = b_i$, $H_{max} = 1$; $0 \le H \le 1$.

For standard experimental conditions, with a pump pulse of 30 ns, $w_i = 0.5$ cm and energy 140 mJ and for an SBS material with the refractive index $n = 1$ and phonon lifetime 0.85 ns, the calculated overall reflectivity reaches 78% and the fidelity factor, 94%, as shown in figure 5.4 (from [5.29]).

Using the numerical calculated data from figure 5.4, denoted by R_{sim} and H_{sim} respectively, one can check the analytical relation between SBS reflectivity and fidelity from equation (5.20). The predictions for the fidelity are denoted by H_t and are shown in table 5.3.

One could remark that the agreement of prediction for fidelity (equation (5.20)) holds well with the numerical calculations even in the non-stationary SBS, for Gaussian pump waves.

The numerical results have shown that the intensity fluctuations in the Stokes pulse produce degradation in the time-resolved and overall fidelity. The simulations with this three-dimensional model confirm that most of the Stokes energy (about 93%) is concentrated in the conjugated mode (i.e. the fundamental mode). However, even for a Gaussian pump, the non-linear propagation in the SBS material yields reflected and transmitted modes, with energy under 5% and smaller energy for higher orders.

In the same particular case of pumping in the fundamental mode, the SBS reflectivity and the fidelity increase exponentially as a function of pump energy and saturate at different levels (figure 5.4), which was previously

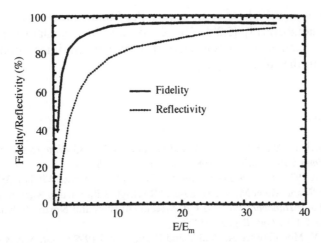

Figure 5.4. Numerical calculation of SBS reflectivity and fidelity ($E_{th} \approx 16\,mJ$ and the other parameters are given in the text) (from [5.29]).

Table 5.3.

$\varepsilon_L/\varepsilon_{L,th}$	2	3	6	8	10	15	20	30
R_{sim}	0.35	0.50	0.70	0.75	0.80	0.86	0.88	0.95
H_{sim}	0.77	0.85	0.92	0.94	0.95	0.97	0.97	0.98
H_t	0.77	0.84	0.96	0.98	0.99	0.99	0.99	0.99

observed and much discussed. Furthermore, the decrease of fidelity was shown at high energy when fast rising pulses (shorter than the phonon lifetime) were used for the SBS pump [5.30].

More generally, if higher order Gauss–Laguerre modes are used for SBS pumping, one can find the same exponential dependence of total reflectivity and fidelity, but growing slower and saturating at lower values for higher modes, as a function of the pump energy. Also, the higher the mode order, the higher the SBS threshold. The decrease of reflectivity for higher modes could be explained by their smaller coupling constants, which lead to smaller gains. The consequence of smaller gains is the decrease of selectivity for the phase conjugated mode, i.e. the possibility of developing other modes which increase or maintain the total reflectivity, but reduce the fidelity. Whenever the SBS total reflectivity is high and the fidelity is low for a certain mode, one can expect that the excess pump energy is coupling that mode to higher modes in the Stokes wave.

Numerical calculations for multi-mode (complex) pump waves show that the fidelity is decreasing for large input spot sizes (short Rayleigh distances), but the reflectivity is practically not affected. In order to ensure high fidelity in the phase conjugation, the Stokes modes should pass the threshold adiabatically [5.29]. This condition could explain why the SBS with broadband lasers is not successful even if the interaction length is shorter than the coherence length.

References

5.1 Zel' dovich B Ya, Pilipetsky N F and Shkunov V V 1985 *Principles of Phase Conjugation* (Berlin: Springer); 1982 *Sov. Phys. Usp.* **25** 713
5.2 Yariv A 1975 *Quantum Electronics* 2nd edition (New York: Wiley) p 387
5.3 Kaiser W and Maier M 1972 Stimulated Rayleigh Brillouin and Raman spectroscopy in *Laser Handbook* vol 2 ed F T Arecchi (Amsterdam: North-Holland) p 1077
5.4 Hon D T 1982 *Opt. Eng.* **21** 252
5.5 Babin V, Mocofanescu A, Vlad V and Damzen M J 1999 *J. Opt. Soc. Am. B* **16** 155; 1995 *Proc. SPIE* **2461** 294
5.6 Tikhonchuk V T, Labaune C and Baladis H A 1996 *Phys. Plasmas* **3** 3777
5.7 Giacone R E and Vu H X 1998 *Phys. Plasmas* **5** 1455
5.8 Labaune C, Baladis H A and Tikhonchuk V T 1997 *Europhysics Lett.* **38** 31
5.9 Lecoeuche V, Segard B and Zemmouri J 1999 *Opt. Commun.* **172** 335
5.10 Vlad V I, Damzen M J, Babin V and Mocofanescu A 2000 *Stimulated Brillouin Scattering* (Bucharest: INOE Press)
5.11 Ridley K D, Jones D C, Cook G and Scott A M 1991 *J. Opt. Soc. Am. B* **8** 2453
5.12 Suni P and Falk J 1986 *J. Opt. Soc. Am. B* **3** 1681
5.13 Miller E J, Skeldon M D and Boyd R W 1989 *Appl. Opt.* **28** 92
5.14 Tang C L 1966 *J. Appl. Phys.* **37** 2945
5.15 Menzel R and Eichler H J 1992 *Phys. Rev. A* **46** 7139
5.16 Kummrow A 1993 *Opt. Commun.* **96** 185
5.17 Moore T R and Boyd R W 1996 *J. Nonlin. Opt. Phys. Mat.* **5** 387; 1998 *J. Mod Opt.* **45** 735
5.18 Stoddart P R, Crowhurst J C, Every A G and Comins J D 1998 *J. Opt. Soc. Am. B* **15** 2481
5.19 Visnyauskas V, Gayjauskas E and Ghinyunas L 1998 Formation of ultrashort optical pulses by induced scattering of light in *Lasers and Ultrashort Processes* p 170 (in Russian)
5.20 Raab V, Heuer A, Schultheiss J, Hodgson N, Kurths J and Menzel R 1999 *Chaos, Solitons and Fractals* **10** 831
5.21 Babin V, Mocofanescu A and Vlad V 2001 *Proc. SPIE* **4430** 533
5.22 Vlad V, Babin V, Mocofanescu A and Eichler H J *Analytical spatio-temporal treatment of stimulated Brillouin scattering* Technical Digest CLEO/Europe-EQEC 2001 Munich 2001 p 250
5.23 Kuzin E A, Petrov M P and Fotiadi A A 1994 Phase conjugation by SMBS in optical fibres in *Optical Phase Conjugation* ed M Gower and D Proch (Berlin: Springer) p 74
5.24 Mashkov V A and Temkin H 1998 *IEEE J. Quantum Electron.* **34** 2036
5.25 Rae S, Bennion I and Cardwell M J 1996 *Opt. Commun.* **123** 611
5.26 Anikeev I Y, Zubarev I G and Mikhailov S I 1986 *Sov. J. Quantum Electron.* **16** 88

5.27 Lehmberg R H 1998 *J. Opt. Soc. Am.* **73** 558

5.28 Hu P H, Goldstone J A and Ma S 1989 *J. Opt. Soc. Am. B* **6** 1813

5.29 Afshaarvahid S and Munch J 2001 *J. Nonlin Opt. Phys. Mat.* **10** 1; 1998 *Phys. Rev. A*
　　　57 3961

5.30 Dane C B, Neumann W A and Hackel L A 1992 *Opt. Lett.* **17** 1271

5.31 Maier M and Renner G 1971 *Opt. Commun.* **3** 301

5.32 Vlad V, Babin V and Mocofanescu A 2002 *J. Optoelectr. Adv. Mat.* **4** 581

Chapter 6

Brillouin-enhanced four-wave mixing (BEFWM)

Stimulated Brillouin scattering (SBS) occurs with great simplicity for high power laser radiation in either a bulk or a waveguide geometry. Under a large range of conditions, both focused or waveguided, the backscattered Stokes wave is found to be predominantly the phase conjugate (wavefront reversed) spatial distribution of the incident light. Unfortunately, the need for threshold requires high input intensity, and for high conversion even higher intensities are desirable. The high intensity requirement limits the practical cases when the SBS process can be used. An additional issue is that the scattered wave is derived from the incident wave energy and the SBS process necessarily is limited to reflectivity less than unity.

An alternative nonlinear optical method of producing reflection and particularly phase conjugation is an interaction scheme known as four-wave mixing (FWM), as illustrated in chapter 4. This, as the name suggests, involves four waves—usually three incident waves and a fourth wave that is generated and is the phase conjugate of one of the incident beams. When this interaction occurs in a Brillouin-active medium and the nonlinearity that couples the fields is the generation of acoustic waves, the process is known as Brillouin-enhanced four-wave mixing (BEFWM).

6.1 The BEFWM interaction geometry

The basic process of BEFWM is a non-degenerate form of FWM, as shown in figure 6.1, in which pairs of the interacting beams (E_1, E_4) and (E_2, E_3) have a frequency separation close to or equal to the resonant acoustic frequency (Ω) associated with the SBS process ($|\omega_1 - \omega_4| = |\omega_2 - \omega_3| = \Omega$). Following the standard terminology of four-wave mixing, we refer to the counter-propagating pair of waves E_1 and E_2 as the pump waves, and the wave E_3 is referred to as the signal (or probe) wave. The pump beams are usually strong waves that supply the optical energy for the interaction, and

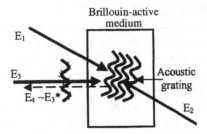

Figure 6.1. Brillouin-enhanced four-wave mixing involving counter-propagating pump beams (E_1 and E_2) and signal (E_3) and conjugate beam (E_4) interaction within a Brillouin-active medium.

the signal beam may be weak. In BEFWM, the signal and the backward pump beams beat together to drive an acoustic wave via electrostriction that modifies the refractive index of the medium. The forward pump beam scatters from this to form the conjugate wave E_4 that is phase conjugate in spatial form to the signal beam E_3.

The first experimental demonstrations of this scheme were made by Basov *et al* [6.1] in order to achieve wavefront reversal of weak pulses in a non-threshold scheme. In addition to the weak probe pulse, a strong pump pulse at the same frequency was used to establish SBS and a resulting counter-propagating Stokes pump pulse from which the probe pulse was reflected in a four-wave mixing type regime. It was shown theoretically and experimentally that the weak probe reflectivity was equal to the reflectivity of the strong pump. One advantage of this scheme is that the wavefronts of the laser and Stokes pumps can be automatically phase conjugated by the normal SBS process and plane pump waves are not required as in normal FWM. The disadvantage of the scheme is that the reflectivity of the probe is limited to a value less than unity due the interaction of the 'pumps', resulting in a depletion of the laser pump E_1.

Decoupling the pump beams to eliminate any interaction between them can increase the efficiency of this process, but the backward pump must be produced outside the four-wave mixing medium. Two methods were used for the decoupling of the pump beams: frequency-decoupled BEFWM and polarization-decoupled BEFWM.

Frequency decoupled BEFWM [6.2, 6.3, 6.4], is accomplished by having the pump beam frequencies non-resonant with the acoustic wave. Many frequency combinations have also been used [6.5, 6.6]. The most common case (and easiest to implement experimentally) is to have the pump wave frequencies the same ($\omega_1 = \omega_2 = \omega$). For the BEFWM, we then need the probe beam frequency detuned from the pumps by the acoustic frequency. One particularly interesting case explored in experiments by Andreev [6.2] is for the probe beam to have anti-Stokes frequency ($\omega_3 = \omega + \Omega$) leading to a generated conjugate wave with frequency $\omega - \Omega$. In experiments, phase conjugate reflectivities $(I_4/I_3) \approx 7 \times 10^5$ were demonstrated. The specific reflectivity is a function of pump intensity and phase mismatch parameter (Δk). In particular, the phase mismatch can result in a phasing between the acoustic contributions from the two resonant interaction terms $E_1 E_4^*$ and $E_2 E_3^*$ to allow a strong temporal instability to be established [6.7] leading to the very high reflectivities observed. The steady-state analysis of the interaction [6.3] also predicts poles in the solutions to the phase conjugate reflectivity in which infinite growth occurs. These solutions neglect the depletion of the pump fields, which is clearly a limitation for the very high reflectivities observed. Scott [6.4] has performed numerical simulations of the transient case with depletion and phase mismatch (Δk) included. A good review paper on the subject was written by Ridley and Scott [6.8].

Polarization decoupled BEFWM occurs when the pump beam polarizations are orthogonal and, owing to the scalar interaction in SBS (in isotropic media), no acoustical coupling between these beams exists. The technique of polarization-decoupled BEFWM was investigated by Efimkov [6.9], Bubis *et al* [6.10], Schroeder *et al* [6.11] and Choi [6.12]. Reflectivities greater than unity were demonstrated and polarization control on the beams established for possible applications to the aberration correction of radiation in optical systems with laser amplifiers.

6.2   Theoretical model of BEFWM

6.2.1   Coupled equations characterizing BEFWM

We consider the interaction geometry of figure 6.2, and assuming the waves are plane and monochromatic, the total optical field can be written

$$E = \sum_{i=1}^{4} \tfrac{1}{2} \mathbf{E}_i(\mathbf{r}, t) \mathbf{e}_i \exp \mathrm{i}(\omega_i t - \mathbf{k}\mathbf{r}) + \text{c.c.} \qquad (6.1)$$

where ω_i and k_i are the optical frequencies and the wavevectors respectively; polarization directions $(\mathbf{e}_i)$ are $\mathbf{E}_i \mathbf{e}_i$. The frequency separations of the waves

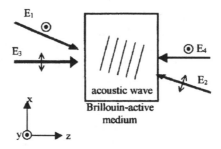

Figure 6.2. Schematic of polarization-decoupled BEFWM.

are: $|\omega_1 - \omega_4| = |\omega_2 - \omega_3| = \Omega$. The fields pairs (E_1, E_4) and (E_2, E_3) yield an acoustical wave with the frequency Ω, described by

$$P = (\tfrac{1}{2})\rho(\mathbf{r}, t) \exp i(\Omega t - \mathbf{k_B r}) + \text{c.c.} \qquad (6.2)$$

where $k_B = \Omega/v$ and v is the acoustical wave velocity.

To describe both BEFWM methods with both pump wave decoupling schemes, one can use a similar set of equations. With frequency decoupling, all fields have the same polarization, and frequencies taken as $\omega_1 = \omega_2 = \omega$, $\omega_3 = \omega + \Omega$ and $\omega_4 = \omega - \Omega$. With polarization decoupling, one can consider: $\omega_1 = \omega_3 = \omega$ and $\omega_2 = \omega_4 = \omega - \Omega$; the beam polarizations are orthogonal: $\mathbf{e}_1 = \mathbf{e}_4$, $\mathbf{e}_2 = \mathbf{e}_3$ and $\mathbf{e}_1\mathbf{e}_2 = 0$; $\mathbf{e}_3\mathbf{e}_4 = 0$.

The equations of BEFWM are simplified by the following (usual) hypotheses of no absorption; neglect of the second-order derivatives with respect to the first-order ones; neglect of the acoustical wave propagation. With these hypotheses, the equation set describing BEFWM can be written as

$$\left(\frac{\mathbf{k}_1 \cdot \nabla}{k_1} + \frac{n}{c} \frac{\partial}{\partial t} \right) E_1 = ig_1 E_4 p$$

$$\left(\frac{\mathbf{k}_2 \cdot \nabla}{k_2} + \frac{n}{c} \frac{\partial}{\partial t} \right) E_2 = ig_2 E_3 p^* \, e^{i\Delta \mathbf{k} \cdot \mathbf{r}}$$

$$\left(\frac{\mathbf{k}_3 \cdot \nabla}{k_3} + \frac{n}{c} \frac{\partial}{\partial t} \right) E_3 = ig_3 E_2 p \, e^{-i\Delta \mathbf{k} \cdot \mathbf{r}} \qquad (6.3)$$

$$\left(\frac{\mathbf{k}_4 \cdot \nabla}{k_4} + \frac{n}{c} \frac{\partial}{\partial t} \right) E_4 = ig_4 E_1 p^*$$

$$\left(\frac{\partial}{\partial t} + \frac{\Gamma_B}{2} + i\Delta\Omega \right) \rho = ig_p (E_1 \cdot E_4^* + E_3 \cdot E_2^* \, e^{i\Delta \mathbf{k} \cdot \mathbf{r}})$$

where g_i is proportional to ω_i, g_p is a constant of the nonlinear coupling, $\Delta \mathbf{k} = \mathbf{k}_1 + \mathbf{k}_2 - \mathbf{k}_3 - \mathbf{k}_4$ is the wavevector mismatch, $\Delta\Omega = v|k_1 - k_4| - \Omega \approx v|k_2 - k_3| - \Omega$ is the acoustic detuning from resonance for scattering at angle θ, ρ is related to the acoustic wave amplitude and Γ_B is the intensity FWHM of the spontaneous acoustic linewidth.

6.3 Polarization-decoupled BEFWM theory

In order to make more explicit the frequencies of the interacting waves, one can introduce the change of notations for the pump fields: $E_1 = E_{L0}$ and $E_2 = E_{S0}$, the probe and the conjugate fields: $E_3 = E_{L1}$ and $E_4 = E_{S1}$, respectively, where the indices L and S correspond to the laser (ω) and Stokes ($\omega - \Omega$) frequencies.

The main amplitude changes occur in the probe and the conjugate beams, so it is easier to take their plane as the x–z plane, and the probe beam direction as the z axis. The polarizations of the fields E_1 and E_4 are parallel and oriented along the y axis and those of the fields E_2 and E_3 are parallel and in the x–z plane. The wavevector Δk is also oriented along the z axis. With these selections, the equation set (6.3) becomes

$$\left(\cos\theta\frac{\partial}{\partial z}+\frac{n}{c}\frac{\partial}{\partial t}\right)E_{L0}=-E_{S1}Q$$

$$\left(-\cos\theta\frac{\partial}{\partial z}+\frac{n}{c}\frac{\partial}{\partial t}\right)E_{S0}=E_{L1}Q^*\,e^{i\Delta kz}$$

$$\left(\frac{\partial}{\partial z}+\frac{n}{c}\frac{\partial}{\partial t}\right)E_{L1}=-E_{S0}Q\,e^{-i\Delta kz} \tag{6.4}$$

$$\left(-\frac{\partial}{\partial z}+\frac{n}{c}\frac{\partial}{\partial t}\right)E_{S1}=E_{L0}Q^*$$

$$\left(\frac{\partial}{\partial t}+\frac{\Gamma_B}{2}+i\Delta\Omega\right)Q=\frac{g_B(\Theta)\Gamma_B}{4}(E_{L0}E_{S1}^*+E_{L1}E_{S0}^*\cos\Theta\,e^{i\Delta kz})$$

where $Q=ig_i\rho$, $\Delta k=(\Delta k)_z=(n\Omega/c)(\cos\theta-1)$ and the frequency detuning $\Delta\Omega=\Omega[\cos(\theta/2)-1]$.

6.3.1 Steady-state and constant pump analysis

A steady-state analysis can be achieved by taking time derivatives in (6.4) to be zero and the acoustic amplitude Q may be directly inserted into the optical field equations. The pump fields E_{L0} and E_{S0} are assumed to be constant in this section and the equations for the weak probe and conjugate fields are described by

$$\frac{\partial E_{L0}^*}{\partial z}=-\lambda(\cos\Theta|E_{S0}|^2E_{L1}^*+E_{L0}^*E_{S0}^*E_{S1}\,e^{i\Delta kz})$$

$$\frac{\partial E_{S1}}{\partial z}=-\lambda(|E_{L0}|^2E_{S1}+\cos\Theta\,E_{L0}E_{S0}E_{L1}^*\,e^{-i\Delta kz}) \tag{6.5}$$

where $\lambda=\frac{1}{2}g_B\cos^2(\theta/2)(1+i\eta)$ is the complex Brillouin-gain coefficient at the angle θ and $\eta=2\Delta\Omega/\Gamma_B$ is the frequency detuning from the acoustical resonance frequency, normalized to the Brillouin linewidth.

The above system of equations can be solved with the boundary conditions $E_{S1}(l)=0$ and the solutions may be written in the form

$$E_{L1}^*(z)=E_{L1}^*(0)\,e^{(p_-+\lambda\cos\theta\,I_{S0})z}\frac{[p_+-p_-\,e^{(p_--p_+)(l-z)}]}{[p_+-p_-\,e^{(p_--p_+)l}]}$$

$$E_{S1}(z)=\lambda\cos\theta\,E_{L0}E_{S0}E_{L1}^*(0)\,e^{(-p_++\lambda I_{L0})z}\frac{[1-e^{(p_--p_+)(l-z)}]}{[p_+-p_-\,e^{(p_--p_+)l}]} \tag{6.6}$$

where $p_\pm=\frac{1}{2}(\mu_c\pm\sqrt{\mu_c^2+4\lambda^2I_{L0}I_{S0}})$ and $\mu_c=\lambda(I_{L0}-\cos I_{S0})+i\Delta k$.

The conjugate reflectivity is given by

$$R = \left| \frac{E_{S1}(0)}{E_{L1}^*(0)} \right|^2 = \left| \lambda \cos \theta \, E_{L0} E_{S0} \frac{[1 - e^{(p_- - p_+)l}]}{[p_+ - p_- \, e^{(p_- - p_+)}]} \right|^2. \tag{6.7}$$

In the simplified case of zero phase mismatch, $\Delta k \approx 0$, $\theta = 0$ and zero detuning ($\eta = 0$), the conjugate reflectivity coefficient becomes

$$R = R_0 = I_{L0} I_{S0} \frac{(1 - e^{\mu L})^2}{(I_{L0} + I_{S0} \, e^{\mu L})^2}. \tag{6.8}$$

In the case of high gain, $e^{\mu L} \gg 1$ and $I_{S0} \, e^{\mu L} \gg I_{L0}$, the conjugate reflection coefficient becomes the ratio of the laser and Stokes beam intensities

$$R = I_{L0}/I_{S0}. \tag{6.9}$$

The presence of frequency detuning and phase mismatch which are often detrimental to FWM can lead to very high reflectivities in BEFWM depending on the corresponding gain parameter $\mu l = \frac{1}{2} g_B (I_{L0} + I_{S0}) l$ and the pump beam ratio R_0.

Figures 6.3 and 6.4 show the dependence of the reflectivity R on the normalized acoustic frequency detuning (η) and, for simplicity, we take $\Delta k = 0$ (Δk is small for most experimental angles, e.g. $\theta = 50$ mrad, $|\Delta k| \approx 10^{-3}$ cm^{-1}).

In figure 6.3 one can observe that the reflectivity rises to a maximum at a value of the detuning which depends on the gain μl. The maximum

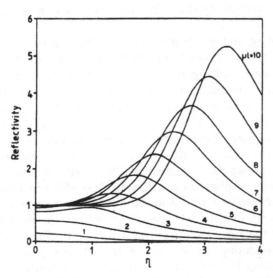

Figure 6.3. Steady-state BEFWM reflectivity (R) against normalized frequency detuning (η) for various pump gain parameters μl and fixed pump ratio $R_0 = 1$ ($\Delta k = 0$).

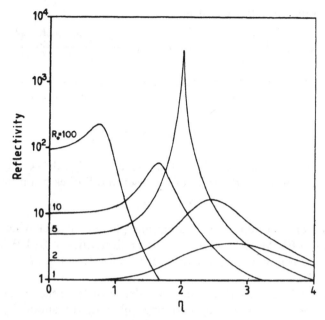

Figure 6.4. Steady-state BEFWM reflectivity (R) against normalized frequency detuning (η) for various pump ratios R_0 and at a fixed gain parameter $\mu l = 8$ ($\Delta k = 0$).

reflectivity is higher than in the case with no detuning, as predicted by equation (6.7). In the graph (figure 6.4) the curve for $R_0 = 5$ shows a pronounced enhancement in reflectivity at $\eta = 2$, since it is close to a pole in the reflectivity equation.

The detuning and phase mismatch can be controlled by the angle θ between the pump and probe beams. If the medium is Kerr-active, intensity dependent refractive index gratings can be established in addition to introducing self-phase modulation to the interacting beams. The additional contribution to the beam coupling by Kerr FWM is, in general, out of phase with BEFWM. To simulate these effects in a real medium we take the experimental parameters for the nonlinear medium, CS_2, which is both strongly Brillouin-active and Kerr-active, at $1.06\,\mu m$ wavelength.

Figure 6.5 shows the calculated BEFWM reflectivity and, to show the relative influences of phase mismatch and Kerr effect in CS_2, three curves are plotted: in curve (a) both Δk and the Kerr effect are neglected, while in curve (b) it is seen that the introduced Δk perturb the peak reflectivity. In curve (c) both the phase mismatch and the Kerr effect are included and there is a considerable suppression of the reflectivity as the normalized frequency-detuning is increased. This is caused by destructive interference between the Kerr term and the imaginary non-resonant component of the Brillouin term.

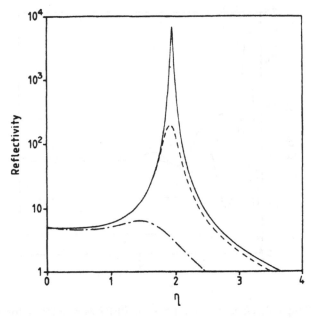

Figure 6.5. Steady-state BEFWM reflectivity (R) against normalized frequency detuning (η) calculated for experimental parameters of CS_2 at $1.06\,\mu m$. Curve (a) (solid line) is with the neglect of Δk and the Kerr effect, curve (b) (dashed line) includes Δk and curve (c) (dot-dashed line) also includes Kerr effect ($\mu l = 8$, $R_0 = 5$).

6.3.2 Transient polarization-decoupled BEFWM

In the BEFWM process, the time taken to establish a steady-state regime can be many times the acoustic decay time (τ_B) [6.9] and for typical laser pulse durations of a few tens of nanoseconds the process is highly transient.

The full transient system of equations (6.4) was numerically solved using a similar scheme to [6.13] in which the optical field equations (6.4) are transformed along their characteristics and a simple numerical difference scheme with adequately small time step used. The accuracy of such a program can be checked by monitoring the photon conservation conditions

$$\int_{-\infty}^{\infty} (|E_{L0}|^2 + |E_{S1}|^2)\,\mathrm{d}z = \text{const.}$$

$$\int_{-\infty}^{\infty} (|E_{S0}|^2 + |E_{L1}|^2)\,\mathrm{d}z = \text{const.} \tag{6.10}$$

In figure 6.6 is shown the temporal variation of the BEFWM reflectivity for various values of the gain parameter μl and a fixed pump ratio ($R_0 = 1$). Phase mismatch and frequency detuning are neglected and the case $\theta = 0$ is taken. The transient reflectivity is noted to overshoot and oscillate towards its final steady-state value, given by equation (6.8). At high gains, the

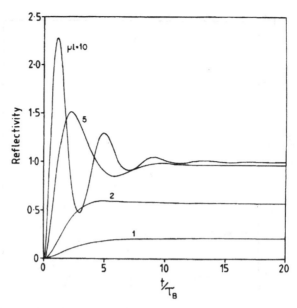

Figure 6.6. Temporal response of the transient BEFWM reflectivity for various gain parameter μl and fixed pump ratio $R_0 = 1$.

transient oscillations are more pronounced and the onset of the first reflectivity peak is faster.

In figure 6.7, the gain parameter is fixed at $\mu l = 5$ and the pump ratio (R_0) varied. It can be established from the results that the temporal growth rate of the BEFWM reflectivity is both a function of the gain parameter and the pump ratio. The numerical model was used to calculate the peak transient reflectivity of a weak probe as a function of gain parameter, and is displayed in figure 6.7. High reflectivities are achieved when the pump ratio is large and for sufficiently high gain, and can be significantly greater than the steady-state value. The results indicate the potential of polarization-decoupled BEFWM for the production of very high reflectivities.

6.4   Experimental investigations on polarization-decoupled BEFWM

An experimental investigation [6.11] into the properties of the polarization-decoupled BEFWM process used a Q-switched Nd:YAG oscillator operating on a single longitudinal and transversal mode. The laser radiation had a pulse duration of 16 ns, energies up to 120 mJ, a beam diameter of ~2 mm, and vertical linear polarization. In the experimental system, shown in figure 6.8, the pump and probe beams for the FWM interaction are derived from the vertically polarized laser using beamsplitters R_1 and R_2.

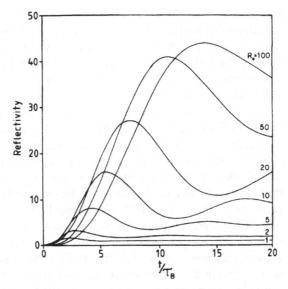

Figure 6.7. Temporal response of the transient BEFWM reflectivity for various pump ratios R_0 and fixed gain parameter $\mu l = 5$.

The laser pump E_{L0} propagates through the FWM cell and into an external SBS cell in which the counter-propagating Stokes pump E_{S0} is generated. This arrangement ensures that the pumps are automatically phase conjugates of each other and means that the pump fields need not be of high spatial quality, as required in normal FWM, for high conjugation

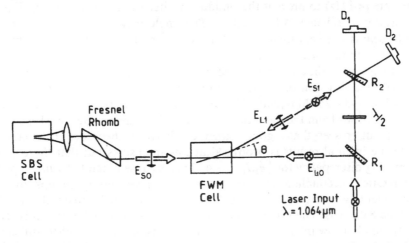

Figure 6.8. Experimental arrangement used in the investigations of polarization-decoupled BEFWM in liquids acetone and CS_2.

fidelity of the probe. A Fresnel rhomb ensures that the counter-propagating beams are of orthogonal polarization. A small fraction of the laser input transmitted by the beamsplitter R_1 passes through a half-wave retardation plate to form the probe beam E_{L1} orthogonal in polarization with E_{L0}. The probe is then injected into the FWM cell with production of an orthogonal polarized Stokes-frequency phase conjugate E_{S1}.

In the experiment the angle θ between the pump and the probe beams is maintained at less than 50 mrad, for a good overlap of the interacting fields. The small angle, together with the use of the same nonlinear medium in both cells, also ensures that the interaction is near the acoustic resonance. In addition the reflectivities of the beamsplitters R_1 and R_2 ensure that the experimental study is conducted in the weak probe case with an intensity pump–probe ratio of $>100:1$. As Brillouin-active liquids, acetone and carbon disulphide (CS_2) were used in the experimental investigation. For the case of CS_2, the laser pump was coupled into a waveguide in the SBS cell in order to maintain the conjugate fidelity. Both arrangements ensured that the process occurred in a regime which was well above its steady-state threshold and provided the necessary conditions for the Stokes pump to be generated with conjugate reflectivities in the range 70–95%. The steady-state Brillouin coefficients of both liquids are relatively large, $g_B = 0.014$ cm/MW for acetone and $g_B = 0.04$ cm/MW for CS_2, at the wavelength of 1.06 μm. The nonlinear gain coefficient was varied by using various lengths of the FWM cell (1, 2 and 5 cm). The phase conjugate reflectivity of the BEFWM process was monitored by recording the powers of the laser probe and its conjugate reflection using photodiodes and a fast oscilloscope with a total rise ~ 1 ns.

The experiments undertaken in acetone were conducted at low repetition rate (<1 Hz) to prevent the buildup of thermal effects due to the linear absorption coefficient of 0.023 cm^{-1}. The angle between the pump and the probe beams was kept small ($\theta = 46$ mrad) to obtain a good overlap between the beams in the BEFWM cell.

In figure 6.9 the conjugate reflectivity for three cells ($l = 1, 2$ and 5 cm) is plotted as a function of the laser pump gain $(\frac{1}{2})g_B I_{L0} l$ for the laser pump intensities ~ 30–250 MW/cm^2. The experimental results are compared with the predictions from the numerical model of the transient BEFWM process.

It can be seen that a good agreement between theoretical calculations and experimental results is produced at low intensities but that there is an increasing discrepancy for $\frac{1}{2}g_B I_{L0} l > 4$. This is caused mainly by incomplete polarization decoupling of the two counter-propagating pumps in the experiment, which results in oscillation of radiation between the FWM and the SBS cells. The rate of increase of the unwanted polarization is relatively small at low intensities, but at high intensities it can lead to sufficient pump interaction to cause depletion of the laser pump field (E_{L0}) and this results in the observed reduction in the BEFWM reflectivity of the probe

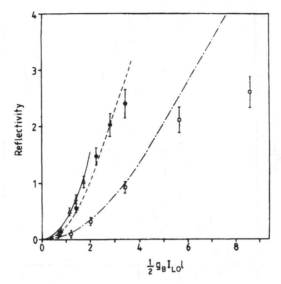

Figure 6.9. Experimentally measured phase-conjugate reflectivity in acetone as a function of gain parameter $\frac{1}{2}g_B I_{L0}l$ for three cell lengths l (triangle: $l = 1$ cm, circle: $l = 2$ cm, and square: $l = 5$ cm).The curves are the theoretical predictions of a numerical solution in the full transient equations.

(E_{L1}) compared with the numerical model which neglected depolarization effects.

The experimental investigation was also carried out using CS_2 as the Brillouin-active medium with the Stokes pump being generated in the external SBS cell with a reflectivity of between 70 and 95%. A rapid increase in the measured phase conjugate reflectivity with the laser pump power parameter $(\frac{1}{2}g_B I_{L0}l)$ at each of the three FWM cell lengths is shown in figure 6.10.

The higher steady-state Brillouin-gain coefficient of CS_2 results in equivalent FWM reflectivities being achieved at lower laser pump intensities. The rate of growth of the 'unwanted' depolarization is also comparable in both experiments, so that depletion of the laser pump in the FWM cell again perturbed the phase conjugate reflectivity as $\frac{1}{2}g_B I_{L0}l$ becomes greater than about 4.

From the results in figures 6.9 and 6.10 it is seen that probe reflectivities in excess of 300% were generated with CS_2 whereas the FWM interaction exhibited lower reflection coefficients at equivalent values of $\frac{1}{2}g_B I_{L0}l$ when acetone was used as the Brillouin medium. One cause of the discrepancy can be accounted for by the effect of the Kerr nonlinearity of CS_2. The Kerr activity contributes an extra coupling term $g_K = 4.6 \times 10^{-3}$ cm/MW to the nonlinear coupling coefficient which is $\pi/2$ out of phase with the

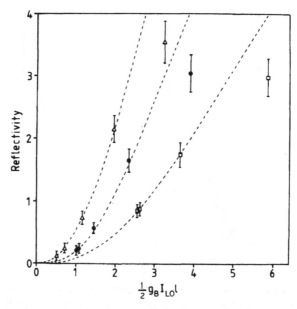

Figure 6.10. Experimentally measured phase conjugate reflectivity in CS_2 as a function of gain parameter $\frac{1}{2}g_B I_{L0} l$ for three cell lengths l (triangle: $l = 1$ cm, circle: $l = 2$ cm, and square: $l = 5$ cm).

steady-state Brillouin-gain coefficient, g_B. At equivalent values of $\frac{1}{2}g_B I_{L0} l$ the FWM process is expected to produce higher phase conjugate reflectivities with CS_2, since the magnitude of the resulting complex nonlinearity is greater than g_B alone. In the experimental system, self focusing of the pump radiation occurred in the FWM cell, which reduced the spot size of the pump laser beam by up to a factor of 2, causing a significant variation in the intensity of the pump beams along the interaction length. The resulting increase in the effective FWM interaction length, together with the increased intensity of the pump beams due to the self-focusing in the FWM cell, contributed to the larger phase conjugate reflectivity observed with CS_2.

In order to avoid complications associated with the intensity-dependent changes in the refractive index and the reduction of the interaction length due to the finite pump–probe angle, the optical alignment shown in figure 6.8 was modified to include a 3 mm diameter, 5 cm long glass capillary waveguide in the FWM cell, to confine the interacting pump and probe fields. This arrangement allowed the pump–probe angle θ to be altered within the limits set by the critical angle for total internal reflection ($\theta_c \approx 320$ mrad) while maintaining fixed values for the interaction length ($l = 5$ cm) and the intensity of the fields in the nonlinear process. Since $|\Delta\Omega| = \Omega(1 - \cos\theta/2)$, the effect of detuning from the acoustic resonance on the transient conjugate reflectivity of the BEFWM process could be investigated experimentally.

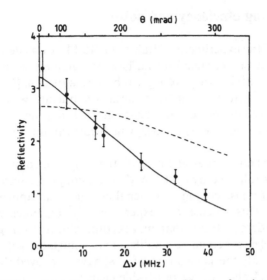

Figure 6.11. BEFWM reflectivity in CS_2 in a waveguide as a function of the frequency detuning produced by varying the probe angle. Solid line is numerical calculation of BEFWM including Kerr nonlinearity of CS_2, dashed line only includes Brillouin.

Figure 6.11 displays the results of monitoring the transient probe reflectivity of the BEFWM interaction as a function of the detuning from the acoustic resonance $\Delta\nu = \Delta\Omega/2\pi$. At a small frequency detuning ($\Delta\nu = 5\,\text{MHz}$) the use of a waveguide produced a reflection coefficient $\sim$320%. As the detuning from resonance was increased by enlarging the interaction angle θ, the conjugate reflectivity was observed to decline to $\sim$25% of its initial value at the critical angle for total reflection θ_c.

At small pump–probe angles ($\theta < 100\,\text{mrad}$) the Kerr nonlinearity increases the effective magnitude of the total FWM coupling coefficient which results in a higher probe reflectivity. As the detuning from the acoustic resonance increases, the increasing imaginary part of the Brillouin-gain, which is π out of phase with the Kerr nonlinearity of CS_2, causes strong destructive interference. Together with the ordinary Lorentzian decrease of the real part of the Brillouin-gain with the detuning $\Delta\nu$, this causes a significant reduction in the magnitude of the complex nonlinearity and, hence, an enhanced reduction in the reflectivity of the FWM process. As a result, the conjugate reflectivity of the nonlinear process continues to decline more rapidly in CS_2 than if the Brillouin liquid were not Kerr-active. This trend is in agreement with that produced by steady-state theoretical analysis where the addition of the Kerr nonlinearity to the Brillouin interaction greatly diminished the reflectivity of the FWM process (figure 6.5) as the normalized detuning η increased beyond a value of $\sim$1.5.

6.5   Scattering efficiency and noise

On the basis of the experiment of Bubis *et al* [6.14] it was demonstrated that BEFWM with a Stokes signal would have the greatest sensitivity for weak signals. This sensitivity was investigated by Bespalov *et al* [6.15] who attenuated the Stokes-shifted signal beam until a conjugate was just detectable above the background noise. The results showed that the minimum resolvable signal has an input energy of 13 fJ, which corresponds to 7×10^4 photons per pulse.

It has been shown by Andreev [6.16] that a system comprising a conventional amplifier followed by a BEFWM phase conjugate mirror, is theoretically capable of phase conjugating signal energies that approach those of a single photon. This scheme was experimentally implemented by Kulagin *et al* [6.17] who demonstrated that images containing an average of 5 photons per resolvable element could be amplified and conjugated with a total gain of 10^{12}. More recently the same authors [6.18] have improved the sensitivity to an average of 2.4 photons per resolvable element.

6.6   Experiments and results in BEFWM for high resolution imaging

The BEFWM process has two important advantages over most conventional imaging: the temporal resolution that comes from using nanosecond laser pulses; and the narrow bandwidth of the BEFWM process, which allows for very good discrimination against background light. One possible application would be imaging of laser material processing, where there would be a lot of broadband and coherent background radiation to discriminate against. This type of application has been suggested by Kulagin *et al* [6.19]. Bespalov *et al* [6.20] have used BEFWM to image a picture through a flame, and have also demonstrated the use of BEFWM to detect scattering from density fluctuations of aerosols in air.

In this section, the investigation of a polarization-decoupled BEFWM configuration used for short-pulse high-resolution imaging will be mainly presented, to illustrate this technique. In this geometry, the four-wave mixing medium is carbon disulphide (CS_2) and the counter-propagating pump is generated from the forward pump by stimulated Brillouin scattering in a second CS_2 cell. This configuration produced a good quality, high-resolution ($\sim$9 μm) phase conjugate image, correcting for aberrations in the system.

The experimental system used is shown in figure 6.12 [6.21]. The input radiation was provided by a Q-switched, frequency-doubled Nd:YAG laser ($\lambda = 532$ nm) operating in a single longitudinal and TEM_{00} mode with a beam diameter of 1.2 mm. The system produced 13 ns pulses (FWHM) at a repetition rate of 10 Hz. The beam from the laser was split into the pump

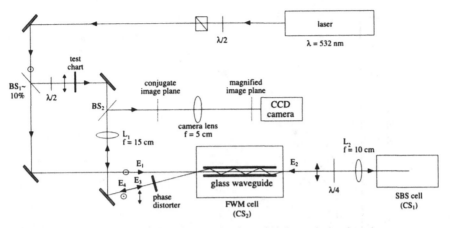

Figure 6.12. Experimental set-up for investigation of high-resolution imaging.

and signal beams at beam-splitter BS_1, which had 10% reflectivity for the vertically polarized laser beam. The pump beam E_1 was directed along the axis of a cylindrical glass waveguide (length 15 cm and inner diameter 2 mm) immersed in the FWM cell containing the Brillouin-active medium CS_2, and then focused by lens L_2 onto the back cell, also containing CS_2. This produced a phase conjugate backward pump E_2 via SBS in the back cell. The quarter-wave plate between the cells ensured the pump beams were orthogonal polarized in the FWM cell.

The signal beam E_3 is passed through a $\lambda/2$ plate making it orthogonal to E_1, but able to interact with the backward pump E_2. The signal is injected into the waveguide at an external angle of $\sim 6°$. Using CS_2, which has a higher refractive index ($n = 1.62$) than glass ($n \approx 1.5$), ensures that total internal reflection occurs in the waveguide. A USAF test chart is placed into the signal beam and lens L_1 (diameter 5 cm) is used to form the image near to the entrance of the waveguide. The primary purpose of this lens imaging is to maximize the coupling efficiency of the signal into the waveguide. There is no requirement for high-quality imaging within the system for it to operate properly. To simulate further aberrations various phase distorters are also placed in the signal beam to test the correction of the system. The quality of the phase conjugate (E_4) of the signal is studied by splitting it at beam-splitter BS_2 and placing a CCD camera at the conjugate image plane to the test chart. Since the resolution of the detected image is limited by the pixel size of the CCD (11 μm × 11 μm), we use a corrected camera lens to magnify the image by ×3 onto the camera.

Initially, a study was made of the characteristics of the system without the test chart in the signal beam. The Stokes' beam, E_2, generated in the back cell, was monitored temporally with a fast photodiode and a digital oscilloscope and its spatial quality with a two-dimensional beam profiler

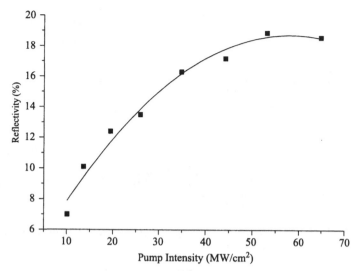

Figure 6.13. Phase conjugate reflectivity against pump intensity for BEFWM.

system with a CCD camera. These showed the Stokes pulse had good Gaussian spatial quality, but a slightly shorter pulse duration (11 ns) compared with the input pump (13 ns). The energy reflectivity of the back cell was between 45 and 60% (depending on the freshness of the CS_2) at an input energy ~ 2 mJ.

The phase conjugate reflectivity of the signal was measured, defined as the energy in the phase conjugate E_4 divided by the input energy of the signal E_3, against the intensity of the pump beam E_1 (with no corrections for Fresnel reflections). The conjugate reflectivity as a function of pump intensity is shown in figure 6.13, showing a maximum reflectivity $\sim 20\%$. The conjugate E_4 typically had a pulse length about half that of the signal E_3.

The results of the image correction experiments are shown in figure 6.14 [6.21]. Thus, figure 6.14(a) shows the phase conjugate image at the CCD camera in the magnified image plane. The corresponding image with a phase distorter in the signal beam is shown in figure 6.14(b). For comparison, figures 6.14(c) and (d) show the image at the camera when an ordinary 100% mirror is placed before the front cell (i.e. the signal does not pass through the waveguide but is reflected straight back). Figure 6.14(c) shows the reflected image with no phase distorter in the system and figure 6.14(d) shows the result after a double pass through the distorter.

The quality of the phase conjugate correction produced by this BEFWM experiment is excellent as can be seen by comparing figures 6.14(a) and 6.14(b). The correction is also very good for the distortion introduced when a stronger phase disturber is used. In the magnified image, the smallest bars that could be resolved are 9 μm (57 line pairs/mm) and this is maintained

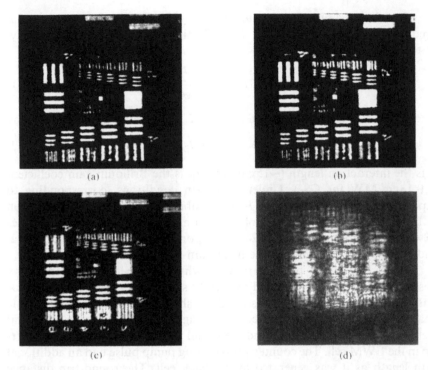

Figure 6.14. (a) Phase conjugate image; (b) phase conjugate image through distorter; (c) image of test chart with mirror; (d) image after double pass through distorter.

in the image with the distorter in the system. The resolution limit of the lens L_1 is calculated to be $\sim3\,\mu m$ (the distance from the lens to the test chart being $\sim30\,cm$).

Theoretically, it is expected that the BEFWM process will have a limit imposed on the resolution due to increasing phase mismatch, as the signal incidence angle θ becomes larger. For CS_2, the angular reduction in efficiency is measured experimentally in [6.11]. This indicates an acceptance angle of $\approx200\,mrad$, which corresponds to a predicted resolution of $\sim2.5\,\mu m$. However, the injected signal entered the waveguide at $\sim100\,mrad$ and the angular spread around this value may contribute to a reduced achievable resolution.

It is seen in figures 6.14(a) and 6.14(b) that the horizontal bars were resolved more clearly than the vertical bars. The reason for this is not clear, but it may be explained by the signal beam entering the waveguide at an angle in a horizontal plane with respect to the pump. This means the signal reflects from side to side whilst propagating down the waveguide but not from top to bottom, which could affect the relative quality of phase conjugation horizontally and vertically. However, this effect is also seen in figure 6.14(c), where the image is reflected back before it enters the waveguide, so this may not be the only explanation.

The observed phase conjugate reflectivity (figure 6.13) is lower than that predicted by steady-state theory. The reflectivity of BEFWM in the steady-state is given in [6.11]

$$R = I_1 I_2 \frac{(1 - e^{\mu L})^2}{(I_1 + I_2 e^{\mu L})^2} \tag{6.11}$$

where I_1 and I_2 are the pump intensities, μL is the gain parameter given by

$$\mu L = \frac{g_B}{2}(I_1 + I_2)L \tag{6.12}$$

L is the interaction length (~ 15 cm) and g_B is the Brillouin-gain coefficient (~ 0.13 cm/MW for CS_2). Under high gain conditions, corresponding to $\exp(\mu L) \gg 1$ and $I_2 \exp(\mu L) \gg I_1$, the probe reflectivity given in equation (6.11) is approximately the ratio of the laser and the Stokes' pump intensities ($R = I_1/I_2$). In our case, $\mu L \sim 20$, corresponding to the high gain criteria, and the pump ratio (and thus the maximum reflectivity obtainable) was 2.

There are several possible reasons why the experimental reflectivity observed is lower than predicted by steady state theory. A principal factor was the transiency of the process. This is shown by the shortening of the phase conjugate pulse with respect to the signal pulse and is due to several factors. The two pump beams and the signal had incomplete temporal overlap in the FWM cell. The counter-propagating pump pulse had an additional path length as it was generated in the back cell. The round-trip distance between the cells corresponds to a delay time of 3 ns, in addition to a delay due to the transient onset of SBS in the back cell. The transience and nonlinearity of the pulsed BEFWM process in the front cell also contribute to the shortening of the conjugate pulse. In BEFWM, the time taken to reach a steady state can be many times (>10) greater than the acoustic decay time τ_B. As $\tau_B \sim 2$ ns in CS_2 [6.11], it can be seen that with pulses of the order of 10 ns the process cannot be considered steady-state. Other contributing factors were the incomplete polarization decoupling of the pump beams and changes in the CS_2, which degrades under illumination (higher reflectivity of up to 45% was observed when the CS_2 was fresh) and leads to increasing thermal effects in the medium [6.20].

References

6.1　Basov N G, Zubarev I G, Kotov A V, Mikhailov S I and Smirnov M 1979 *Sov. J. Quantum Electron.* **9** 237

6.2　Andreev N F, Bespalov V I, Kiselev A M, Matreev A Z, Pasmanik G A and Shilov A A 1980 *JETP Lett.* **32** 625

6.3　Scott A M 1983 *Opt. Comm* **45** 127

6.4　Scott A M and Hazell M 1986 *IEEE J. Quantum Electron.* **QE-22** 1248

6.5　Skeldon M D, Narum P and Boyd R W 1987 *Opt. Lett.* **12** 343

6.6 Bespalov V I, Betin A A, Dyaffov A I, Kulgria S N, Manishin V G, Pasmanik G A and Shilov A A 1980 *Sov. Phys. JETP* **52** 190

6.7 Shilov A A 1982 *Sov. Phys. JETP* **55** 612

6.8 Ridley K D and Scott A M 1994 Brillouin-induced four-wave mixing in *Optical Phase Conjugation* ed M Gower and D Proch (Berlin: Springer)

6.9 Efimkov V F, Zubarev I G, Mikhailov S I, Smirnov M G and Sobolev V B 1984 *Sov. J. Quantum Electron.* **14** 209

6.10 Bubis E I, Pasmanik G A and Shilov A A 1983 *Sov. J. Quantum Electron.* **13** 971

6.11 Schroeder W A, Damzen M J and Hutchinson M H R 1989 *IEEE J. Quantum Electron.* **QE-25** 460

6.12 Choi B I and Nam C H 1999 *Appl. Phys. B* **69** 55

6.13 Damzen M J and Hutchinson M H R 1983 *IEEE J. Quantum Electron.* **QE-19** 7

6.14 Bubis E I, Kulagin O V, Pasmanik G A and Shilov A A 1984 *Sov. J. Quantum Electron.* **14** 815

6.15 Bespalov V I, Matveev A Z and Pasmanik G A 1986 *Radio Phys. Quantum Electron.* **29** 818

6.16 Andreev N F, Bespalov V I, Dvoretsky M A and Pasmanik G A 1989 *IEEE J. Quantum Electron.* **QE-25** 346

6.17 Kulagin O V, Pasmanik G A and Shilov A A 1990 *Sov. J. Quantum Electron.* **20** 292

6.18 Kulagin O V, Pasmanik G A, Potlov P B and Shilov A A 1990 *Sov. J. Quantum Electron.* **20** 1395

6.19 Kulagin O V, Pasmanik G A, Potlov P B and Shilov A A 1989 *Sov. J. Quantum Electron.* **19** 902

6.20 Bespalov V I, Kulagin O V, Makarov A I, Pasmanik G A, Potjomkin A K, Potlov P B and Shilov A A 1991 *Opt. Acoust. Rev.* **1** 71

6.21 Mocofanescu A, Corner L, Garcia R and Damzen M J 1997 *J. Mod Opt.* **44** 731

Chapter 7

Techniques for enhancement of SBS

The process of SBS works most effectively at moderately high peak power ($\sim$MW) with pulsed operation ($\sim$10 ns), corresponding to pulse energies $\sim$10–100 mJ. At lower powers the process will be below threshold and the efficiency is too low to be practical in most laser applications. Even above threshold the efficiency may still not be as high as required. Only at many-times threshold power does the reflectivity of the process reach near unity. At very high intensities, however, the SBS process can also be disrupted by detrimental competing nonlinearities (e.g. Kerr self-focusing, thermally-induced defocusing and SRS) and also degradation of the SBS material (e.g. light induced breakdown, thermally-induced convection, bubble formation). A further issue is the need to achieve good quality phase conjugation, requiring selection of the conjugate from noise components, which is also negatively impacted by competing nonlinearities. A still further issue is the transiency of the SBS process that dictates the range of pulse duration that can be used. The net result is that SBS may only operate over a narrow range of laser powers and pulse durations for a given SBS medium and interaction geometry.

Considerable work has been done to try to extend the conditions under which SBS operates efficiently and with good quality of phase conjugation. At one end, techniques have been devised to reduce the threshold and enhance the efficiency of SBS for low powers, at the other end to extend the operation to very high powers. In the previous chapter BEFWM was described as a method of threshold-free SBS. It has, however, the drawback of requiring high power secondary beams to achieve the phase conjugate reflection of a weak beam. In this chapter we look at techniques to enhance SBS involving the input of a single beam required to be phase conjugated. For threshold reduction, the technique of optical feedback is described. At the high power end, the use of two-cell SBS is described and also coherent beam-combining. Finally, geometries are described to achieve laser pulse compression with high efficiency.

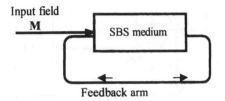

Figure 7.1. A schematic diagram illustrating the principle of the optical feedback technique.

7.1 Optical feedback used to enhance stimulated scattering

Optical feedback is a method to improve the operational characteristics of the phase conjugation process of stimulated Brillouin scattering (SBS). The technique has been shown to reduce the power requirement and enhance the reflectivity of the SBS process, and also provide means of obtaining efficient, high-fidelity phase conjugation of high-energy laser beams.

A general schematic diagram illustrating the principle of the optical feedback technique with SBS is depicted in figure 7.1.

A pump beam is injected into a Brillouin-active medium where, if it is sufficiently intense, it can generate a Stokes-reflected signal from the acoustic wave generated in the SBS process. Optical feedback is accomplished by taking the pump output from the back of the SBS cell and redirecting it to overlap with the main pump input through the front of the SBS cell. The feedback beam can be combined with the main beam via, for instance, a beamsplitter or overlapped with an angular offset from the main beam. The feedback of the pump not only directly enhances the acoustic wave generation but, more importantly, the Stokes radiation, retracing the path of the feedback pump, acts as a seeding field at the back of the SBS cell. Hence, as far as the main pump beam is concerned, the scattering process is no longer initiated from spontaneous noise but rather from a comparatively large Stokes-feedback seeding signal. The consequence of the feedback is a significant enhancement of the SBS process leading to a reduction of the threshold power requirement of the process [7.1, 7.2] and higher reflectivity.

In experiments, two main feedback geometries have been used [7.1,7.2]—in one case a beamsplitter was used as the feedback mechanism, effectively forming a ring cavity, and another case involved injecting the feedback beam at a small angle to the main input. We describe these two cases separately.

7.2 Optical feedback from a beamsplitter (ring resonator)

Input radiation with intensity $|M|^2$ is passed through a beamsplitter of amplitude reflectivity r and is injected into the SBS cell, as shown in figure 7.2, to

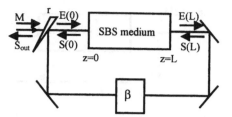

Figure 7.2. A schematic diagram of the configuration of optical feedback using a beam-splitter. An amplitude transmission factor (β) of the feedback arm is included.

form a clockwise circulating laser field E, and through SBS, an anticlockwise Stokes field S is also established in the ring cavity.

An amplitude transmission factor β is introduced to account for optical losses around the feedback loop (excluding the beamsplitter losses). Losses at both the laser and Stokes frequencies are taken to be the same and phase accumulation due to the optical path length of the feedback arm can be omitted (for simplicity) by taking the round-trip time of the feedback cavity to be resonant at both laser and Stokes frequencies. The resonance condition of the feedback cavity for the Stokes frequency, assuming resonant laser frequency, is $\tau = 2m\pi/\omega_{\mathrm{B}}$, where τ is the round-trip time of the ring cavity, ω_{B} is the acoustic angular frequency, and m is an integer.

In this geometry, the normal SBS equations apply

$$\frac{\mathrm{d}S}{\mathrm{d}z} = -\frac{g}{2}|E|^2 S, \qquad \frac{\mathrm{d}E}{\mathrm{d}z} = -\frac{g}{2}|S|^2 E \tag{7.1}$$

but now with different boundary conditions

$$E(0) = M\sqrt{(1-r^2)} + r\beta E(L), \qquad S(L) = r\beta S(0) \tag{7.2}$$

where g is the steady-state Brillouin-gain coefficient. Plane waves are assumed in this analysis and the laser is taken to enter the front of the SBS cell ($z = 0$) with the back of the cell at $z = L$. At the front of the SBS cell the laser field $E(z = 0)$ is the combination of the fraction of the input pump field M transmitted by the beamsplitter and the feedback of the laser throughput from the back of the SBS cell. At the back of the cell, instead of the normal noise level $\approx |E|^2 \mathrm{e}^{-30}$ that initiates the SBS interaction without feedback, the seeding Stokes signal $S(L)$ is a significant fraction of the output Stokes field $S(0)$. With the two boundary conditions described above the SBS system of equations can be solved [7.2]. A relationship between the overall reflectivity R (the reflectivity observed outside the cavity) and the input pump intensity $|M|^2$ is given by the following set of equations:

$$g|M|^2 L = \frac{(1 - \Psi r\beta)^2 \ln(\Psi/r\beta)^2}{(1 - |\eta|^2)(1 - r^2)} \tag{7.3}$$

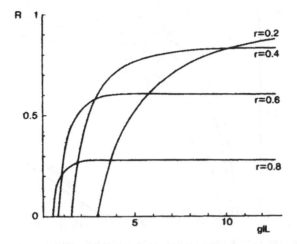

Figure 7.3. The reflectivity R (measured outside the cavity) plotted against the input gain parameter gIL, where $I = |M|^2$ is the input intensity, for different amplitude reflectivities r of the beamsplitter. The transmission factor is $|\beta|^2 = 0.5$.

where $|\eta|^2 = |S(0)|^2/|E(0)|^2$ is the internal SBS reflectivity (as seen by the SBS cell) and $\Psi = \sqrt{[1 - |\eta|^2(1 - r^2\beta^2)]}$.

$$R = |\eta|^2 \frac{(1 - r^2)^2}{(1 - \Psi r\beta)}. \tag{7.4}$$

The threshold input intensity $|M_{th}|^2$ for this case is given by

$$g|M_{th}|^2 L = \frac{(1 - r\beta)^2 \ln(1/r^2\beta^2)}{(1 - r^2)}. \tag{7.5}$$

For example, when $r = 0.2$ and $\beta^2 = 0.5$, a value of $g|M_{th}|^2 L = 3$ is achieved compared with the usual threshold value $\sim$30, representing a 10 times reduction of the threshold power required.

A set of curves of the reflectivity R against $g|M_{th}|^2 L$ is shown in figure 7.3 for different amplitude reflectivities r of the beamsplitter, based on the above equations. It can be seen that the using of a beamsplitter enhances the process by acting as a feedback mechanism, but on the other hand it rejects some of the input pump from injecting the cavity.

The effect of varying the feedback losses (and thus varying the transmission factor β of the feedback arm) is shown in figure 7.4. The observed variation of the reflectivity curves with β is typical for all feedback systems. It is seen from figure 7.4 that although the efficiency of the feedback system is reduced with decreasing β, it is largely insensitive to the losses in the feedback arm. Its performance, therefore, is relatively unaffected over a wide range of losses under steady-state conditions.

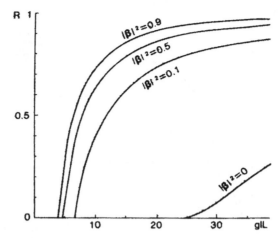

Figure 7.4. The reflectivity R (measured outside the cavity) plotted against the input gain parameter gIL, where $I = |M|^2$ is the input intensity, for different transmission factors $|\beta|^2$ of the feedback arm.

In an experimental investigation of this feedback system [7.2], a single surface from a wedged glass slide placed at $45°$ was used as a feedback beam-splitter, as shown in figure 7.2. At this angle, the reflectivity of the glass slide for the input horizontal polarization was approximately 1%, thus providing only a small level of feedback. The transmission factor $|\beta|^2$ was measured to be approximately 0.5. The output reflectivity R was measured as the ratio of the peak of the Stokes power to the peak of the laser power. In the steady-state the peak-to-peak reflectivity would be equivalent to the energy reflectivity, but in the transient regime, the energy reflectivity would depend on many parameters, such as the input pump intensity and pulse duration, the response time of the SBS medium, the feedback round-trip time, as well as the loss factor of the feedback arm. The laser pulse was provided by a Q-switched frequency-doubled Nd:YAG laser (532 nm) operating on a single longitudinal and transverse mode (TEM$_{00}$). It had FWHM duration of 12 ns, energy varied from 0.3 to 1.1 mJ and was focused onto a 5 cm long Brillouin cell containing n-hexane. The input gain parameter $g|M|^2L$ was deduced for a focused beam taking the Gaussian beam parameters $|M|^2 = P/\pi w_0^2$ and $L = b = k w_0^2$ where P is the peak of the laser power, w_0 is the focal waist size, b is the confocal beam parameter, and k is the wavevector of the radiation in the medium. This gives the result $g|M|^2L = 2gP/\lambda$, where λ is the pump wavelength and is independent of the focal length used with $g = 0.2$ cm/MW and $\lambda = 532$ nm.

The experimental results for the reflectivity of the system for various input laser powers are displayed graphically in figure 7.5. The results show that the reflectivity and the threshold were significantly enhanced when the feedback arm was introduced, even at the low feedback factor ($r\beta = .07$)

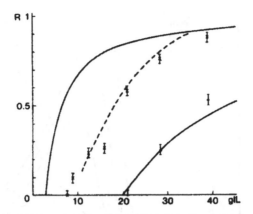

Figure 7.5. A graph of the overall reflectivity R (measured outside the cavity) against the input parameter gIL (with $I = |M|^2$, $|\beta|^2 = 0.5$ and $|r|^2 = 1\%$). Symbols denote experimental results with optical feedback ($\times$) and without feedback ($\bullet$). The leftmost solid curve is the steady-state theoretical prediction with feedback and rightmost curve without feedback. The broken curve is a transient numerical result of a Gaussian pulse with 12 ns FWHM.

used in this experiment. It is seen that at a moderate gain value of $g|M|^2L = 40$, a high reflectivity of 90% is achieved. The discrepancy between the experimental points and the steady-state predictions is attributed to transient effects. This was confirmed by a numerical simulation of the experiment which takes account of the transiency of the SBS process and the finite propagation time around the feedback loop [7.2]. The results are shown as the broken curve in figure 7.5. Good agreement is found between the experimental results and the numerical simulations, verifying the importance of the transiency in the experiment.

It has also been shown numerically that, when the laser intensity is high enough and the feedback losses small, a transient overshoot in reflectivity above the steady-state value can occur such that greater than 100% (peak-to-peak) reflectivity can be achieved. Figure 7.6 shows a numerical simulation of the Stokes signal for a 12 ns input pulse and the occurrence of greater than 100% reflectivity with a high transmission factor β.

The greater-than-unity reflectivity can be explained by the storage of radiation in the feedback system. Since the Stokes seeding signal can become a significant fraction of the laser input and is delayed by the round-trip time of the device, the Stokes output can become greater than the input pump. This overshoot in reflectivity above the steady-state value is obviously temporary since the excess depletion of the pump beam will subsequently reduce the level of feedback, producing a transient overshoot peak for a time of the order of twice the round-trip time. The amount of overshoot is dependent on the excess of pump above threshold and also losses in the feedback arm. When the losses are low, the intracavity Stokes seeding can be high and the overshoot can be significant.

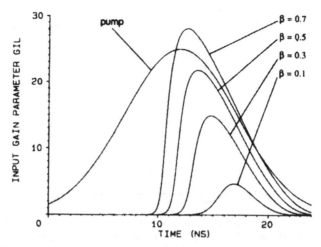

Figure 7.6. Results of numerical simulation of the Stokes signal for various feedback losses $|\beta|^2$ using a Gaussian pulse of 12 ns FWHM.

7.3   Optical feedback using angular offset

This geometry was first tried by Odinstov *et al* [4.1] using a 30 ns input pulse in which they achieved a twelve-fold reduction in threshold. The geometry, incorporating a waveguide in this case, is shown in figure 7.7.

The input pump $E_1(0)$, after passing through the waveguide, is fed to the front of the waveguide $E_2(0)$ where it is injected at a small angle to the input pump. As well as acting like a second pump, enhancing the acoustic interaction, E_2 will also act like a probe beam as in FWM. In this way, there is a Brillouin-enhanced FWM interaction as well as the normal Brillouin scattering occurring simultaneously inside the cell. The effect of this is to have energy transferred from the main pump E_1 to the seeding Stokes field S_2, through the grating set up by E_2 and S_1. This interaction in the SBS

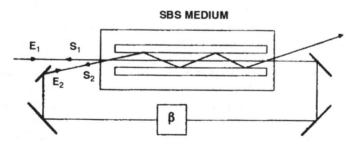

Figure 7.7. A schematic diagram of the configuration of optical feedback using angular offset in a waveguide geometry.

medium is of the type known as non-threshold SBS [7.4, 7.5]. Depending on the strength of the FWM interaction, the seeding level S_2 can be high. Thus as the FWM interaction builds up, a strong seeding signal can be excited and will lead to high SBS reflectivity.

These interactions can be formulated using the SBS equations, including the transverse derivatives, to allow for non-plane waves and non-colinear geometry:

$$-\frac{i}{2k}\nabla_\perp^2 E + \frac{dE}{dz} = -\frac{g}{2}|S|^2 E$$
$$\frac{i}{2k_s}\nabla_\perp^2 S + \frac{dS}{dz} = -\frac{g}{2}|E|^2 S$$

(7.6)

with $E(r) = \sum_{j=1}^2 E_j \, e^{i(kz-\omega t)} e_j(r)$ and $S(r) = \sum_{j=1}^2 S_j \, e^{i(k_s z-\omega_s t)} s_j(r)$, where $e_j(r)$ and $s_j(r)$ are the spatial forms of the laser Stokes components and E_j and S_j are their mean amplitudes. This system of equations has been solved numerically in [7.6].

In one experiment [7.2], the input pump was telescoped down in size to pass through a waveguide consisting of a 2 mm diameter and 6 cm long glass capillary tube immersed in CS_2 liquid. Liquid CS_2 has a sufficiently high refractive index ($n = 1.62$) compared with glass ($n = 1.5$) to allow total internal reflection for waveguiding the light. The throughput was recollimated by a lens, and then redirected with prisms to the front of the waveguide and re-injected at a small angle (about 5°) to the main pump. The losses in the feedback arm were approximately 50%. The reflectivity of this feedback system is shown in figure 7.8. It is seen that while the system shows a large enhancement of reflectivity, it falls short of the reflectivities predicted by the numerical simulations.

It was found that, in the experiment, the waveguide was somewhat under-filled due to self-focusing in the CS_2 and consequently the overlap of the feedback pump and the input pump was less than perfect. The computer simulation, which assumes a perfect overlapping of the two beams, over-estimates the reflectivity and this effect is shown in the discrepancy between the numerical simulations and the experimental results.

A modified version of the above experiment was done, using a focused geometry instead of a waveguide (figure 7.9).

In this case a more careful consideration of the overlap of the feedback beam and the main pump beam needed to be made. As long as the feedback pump intersects at the front region where the acoustic was strongest, appreciable seeding would be generated. However, such a situation did mean that the gain length of the overlap interaction was reduced, so this system was slightly less effective than the waveguide case. Nevertheless, as the results in figure 7.10 indicate, it was still effective in increasing the efficiency of the SBS process.

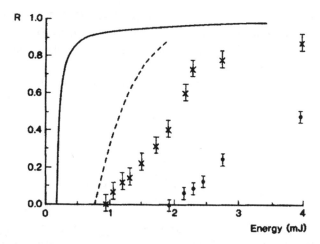

Figure 7.8. The reflectivity R (measured outside the cavity) against the input pump energy. ($\times$) denotes the experimental results with optical feedback; ($\bullet$) denotes the experimental results without feedback. The solid curve is the steady-state theoretical prediction with feedback, and the broken curve is the transient numerical result of a Gaussian pulse with 12 ns FWHM ($|\beta|^2 = 0.5$).

The quality of phase conjugation has been also investigated in this system. By passing the TEM_{00} beam through a phase plate the divergence was increased to approximately 20 mrad, which was about 50 times the diffraction-limited divergence. Using feedback with an angular offset and focused beam geometry, the phase conjugate fidelity without feedback was within the range 60–75% when the total Stokes reflectivity varied from 30 to 65%. The phase conjugate fidelity was obtained by measuring the percentage of the Stokes radiation returning within the solid angle of the input pump after double passing the phase plate. When feedback was included the fidelity was increased to the range 80–90 % when the total Stokes reflectivity varied from 30 to 90 %.

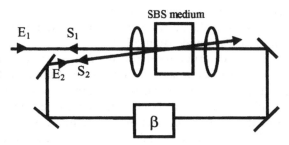

Figure 7.9. A schematic diagram of the configuration of optical feedback using angular offset in a focused geometry.

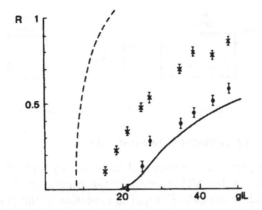

Figure 7.10. The reflectivity R (measured outside the cavity) against the input parameter gIL (where $I = |M|^2$ and $|\beta|^2 = 0.5$). Symbols denote the experimental results with optical feedback ($\times$) and without feedback ($\bullet$). The solid curve is the steady-state theoretical prediction without feedback, and the broken curve is the transient numerical result of a Gaussian pulse with 12 ns FWHM (with feedback).

7.4 Two-cell SBS system

The process of SBS can be efficient for phase conjugating pulsed laser radiation at moderate powers (a few megawatts) and high-fidelity phase conjugation can be achieved using a focused geometry or by propagating through a waveguide containing a Brillouin-active medium. At higher laser power the onset of competing nonlinearities such as optical breakdown, thermal heating, and self-focusing can completely disrupt the SBS process, reducing efficiency and phase conjugation fidelity. The problem of high intensity can be overcome by scaling to larger-aperture geometries, but the phase conjugate fidelity is severely degraded owing to the increased number of uncorrelated Stokes modes that are amplified [7.7].

A simple method for extending the dynamic power range of the SBS mirror is by using a two-cell SBS system. A schematic of the two-cell configuration is shown in figure 7.11. The arrangement consists of a large-aperture front cell (SBS amplifier) that receives the full incident power (P_a) in series with a small-aperture back-cell (SBS generator) that receives a fraction of the incident power (P_g) owing to pump depletion in the SBS amplifier and any intercell attenuation. The generator cell can be arranged in a geometry appropriate for obtaining high-fidelity phase conjugation, such as one that uses a small-diameter waveguide or tight focusing.

In this configuration the amplifier cell is seeded by a phase conjugate Stokes beam (P_g^s) rather than being initiated from spontaneous noise. Several authors have reported efficient phase conjugation of high power, near diffraction-limited radiation using the generator–amplifier scheme [7.8–7.10].

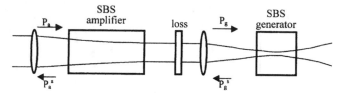

Figure 7.11. A schematic diagram of the configuration of a two-cell SBS system.

7.4.1 Theoretical predictions of two-cell SBS

In order to compare the experimental situation with theory, the transient equations describing the SBS process in the generator and amplifier cells has been modeled numerically [7.13]. The evolution of the amplifier (generator) pump field E_a (E_g), Stokes field S_a (S_g), and acoustic-wave amplitude Q (Q_s) are described by the following system of equations:

$$\left(\frac{\partial}{\partial z}+\frac{n}{c}\frac{\partial}{\partial t}\right)E_{a,g}=-\frac{g_B}{2}\Gamma S_{a,g}Q_{a,g}$$

$$\left(\frac{\partial}{\partial z}-\frac{n}{c}\frac{\partial}{\partial t}\right)S_{a,g}=-\frac{g_B}{2}\Gamma E_{a,g}Q_{a,g}^* \tag{7.7}$$

$$\frac{\partial Q_{a,g}}{\partial t}+\Gamma Q_{a,g}=E_{a,g}S_{a,g}^*$$

where c/n is the speed of light in the medium, g_B is the steady-state Brillouin-gain coefficient, and $\Gamma_B = 2\Gamma$ is the Brillouin linewidth. Plane waves are assumed in the simulation, but the non-collimated geometry is accounted for by allowing an area variation $A_{a,g}(z)$ which, together with the laser power in the two cells $P_{a,g}$, determines the intensity $I_{a,g}$:

$$I_{a,g}(z,t)=\frac{P_{a,g}(z,t)}{A_{a,g}(z,t)}=|E_{a,g}(z,t)|^2. \tag{7.8}$$

The boundary conditions appropriate to the two-cell system are

$$P_g(0,t)=TP_a(L_a,t-\tau)$$

$$P_a^s(L_a,t)=TP_g^s(0,t-\tau) \tag{7.9}$$

for laser powers $P_{a,g}$ and Stokes powers $P_{a,g}^s$ where T is the intercell transmission factor and τ is the intercell time of flight. For comparison with experimental work, the input-pump beam to the amplifier is specified as a Gaussian pulse with 12 ns FWHM duration and the SBS interaction is initiated by the usual spontaneous noise. The numerical calculations of the temporal forms of the pump input and Stokes output from the amplifier and generator cells are reproduced in figure 7.12 with a steady-state calculation superimposed, which gives a qualitative guide to the average behaviour of the whole system (figure 7.12(b)–(d)).

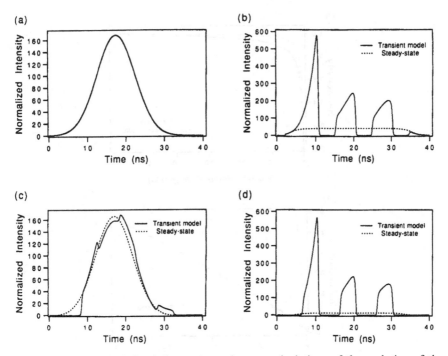

Figure 7.12. Numerical simulation and steady-state calculations of the evolution of the pump radiation into (a) the amplifier and (b) the generator and of the Stokes radiation out of (c) the amplifier and (d) the generator.

The steady-state behaviour of the intercell radiation rapidly deviates from the average of the transient behaviour as the input energy is increased. The transient and steady-state reflectivity (R) and amplification factors (A) have also been calculated and are plotted against pulse energy in figure 7.13. As can be seen in figure 7.13(b) the steady-state behaviour of the amplification deviates from the transient model owing to its sensitive dependence on the level of intercell Stokes radiation. This behaviour is consistent when compared with experimental results [7.13] which show a reasonable agreement with the transient modelling both in the qualitative trends and in the actual data.

7.4.2 Experimental results

An experimental arrangement of the two-cell SBS system is shown in figure 7.14 [7.13]. Laser radiation was provided by the frequency-doubled (532 nm), Q-switched output from a Nd:YAG oscillator–amplifier chain operating on a single-longitudinal and single transversal mode (TEM_{00}).

The pulse energy was varied up to a maximum of approximately 100 mJ with a pulse duration of 12 ns FWHM. Temporal data were recorded using

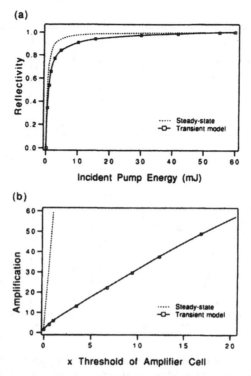

Figure 7.13. Transient model (square points) and steady-state calculation against energy of (a) reflectivity (R) and (b) energy amplification factor A.

photodiodes with a rise time of 125 ps and transient digitizing oscilloscopes with a bandwidth of 250 MHz. The SBS generator–amplifier system shown in figure 7.14 consists of two cells arranged in a series, each preceded by a converging lens. The pump radiation enters the system through a long focal length lens ($f = 2.2$ m) situated 1.6 m from the entrance window of

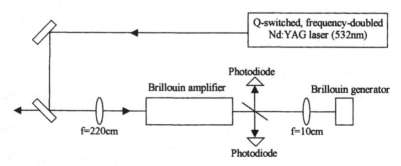

Figure 7.14. Experimental arrangement for studying the two-cell geometry.

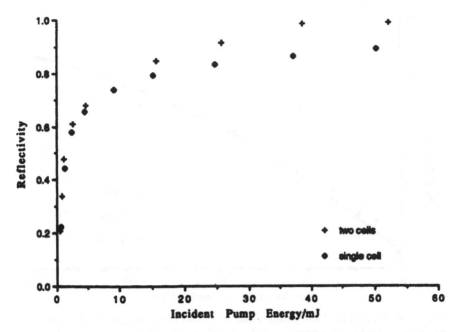

Figure 7.15. Experimental reflectivity (R) for the two-cell system and for a single-cell against input-pulse energy.

the amplifier cell. This arrangement provides soft focusing of the pump radiation and allows control of the intensity of the pump radiation in the amplifier cell by the lens-to-cell separation. A more powerful lens ($f = 10\,\text{cm}$) then focuses the beam onto the generator cell. The separation between the back of the amplifier cell and the front of the generator cell was 25 cm. The Brillouin-active medium was acetone. The generator cell length was 19 cm and different amplifier cell lengths between 2 and 30.5 cm were investigated.

As expected the presence of the amplifier cell suppressed the problem of breakdown up to the maximum available laser energy of 100 mJ. Consequently, as the results of figure 7.15 show, greater reflectivities were obtained using the two-cell system (96%) than by using a single-cell system (90%).

The performance of the system can be also characterized by an energy-amplification factor defined as the ratio of the total output Stokes energy from the two-cell system and the Stokes seed energy entering the amplifier cell from the generator cell. Figure 7.16 shows the amplification (A) measured experimentally for different cell lengths with the input-pump energy normalized to the observational-threshold energy for SBS to occur in the amplifier cell alone. The data demonstrate a smooth trend towards greater amplification at higher pump energies. As expected, the threshold energy of the amplifier equated to the usual exponential gain coefficient (gIl) of approximately 25–30.

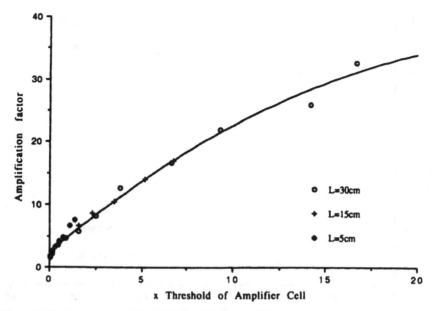

Figure 7.16. Experimental amplification factor (A) for different amplifier-cell lengths contrasted against input-pulse energy normalized to the threshold energy of the amplifier cell.

Further insight into the operation of the two-cell system can be gained by monitoring the temporal evolution of the intercell radiation. An example of the typical behaviour observed for the temporally Gaussian input pulse is shown in figure 7.17. At higher powers the seed pulse (P_g^s) displays some modulation (figure 7.17(b)) that was also present on the incoming intercell pump pulse (P_g). This is due to the sensitive dependence of radiation transmitted through the amplifier on small changes in reflectivity when the amplifier is operated in a highly saturated regime. The transmitted radiation and subsequent seed signal grow and decay periodically, having a behaviour analogous to the relaxation oscillations characteristic of some laser systems. The dependence of the modulation period on the round-trip time was verified experimentally by varying the cell separation. It was found that the modulation period is not merely dependent on the intercell round-trip time but is also a function of the decay and buildup of the acoustic waves in both the generator and amplifier cells. In contrast, the transmitted Stokes signal (P_a^s) of figure 7.17(a) displays little modulation. The oscillatory temporal features of the seed signal are smoothed out owing to the strong saturation of the Brillouin amplifier.

The phase conjugate fidelity of the two-cell backscattered radiation is also of interest. The presence of a seed signal that is well correlated to the input Gaussian beam results in a greater discrimination in the amplifier against non-conjugate light distributions. In the experiments it was found

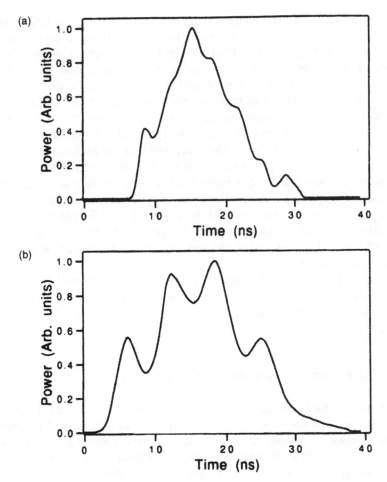

Figure 7.17. Experimental temporal profiles for (a) total reflected Stokes signal and (b) intercell Stokes seed signal.

that with all the pump energies investigated, the two-cell system produced a high quality Stokes output in which virtually all the energy was contained within a small divergence beam. Consequently, the output radiation displayed a high fidelity similar to that measured for a single cell in a focused geometry [7.12].

The generator–amplifier system described above has been demonstrated to extend the use of SBS phase conjugate mirrors to high powers. More detailed considerations of the operation of the system including its temporal and spatial characteristics are given in [7.13]. The physical length and hence the round-trip transit time of the two-cell configuration is an important factor affecting its performance. In order for the two-cell system to attain steady-state behaviour with pulsed laser radiation it is necessary for the pulse duration to be long compared with the SBS onset time and intercell transit time. Similarly, for a coherent interaction it is necessary for the laser-coherence length to be at least twice the total length of the two-cell system. If the intercell transit time is long, then there will be a lengthy delay before the amplifier is seeded and power limiting becomes effective. For this reason it is desirable to use a relatively compact two-cell geometry.

A further feature demonstrated by the simulations but also observed experimentally at high powers is the ability to achieve greater than unity reflectivity on a transient basis. This overshoot by the Stokes radiation can be explained by storage of radiation in the two-cell system. The excess of Stokes power over the input pump will only be temporary since the severe depletion of the pump will reduce the seeding signal, resulting in an over-shoot for a time of the order of twice the intercell transit time. The amount of overshoot is dependent on the gain of the amplifier and on the intercell losses. For low losses, the seeding level can be high, leading to a substantial overshoot.

The performance of the generator cell could be improved by the intro-duction of optical feedback, which has been shown to reduce the threshold and increase the phase conjugate fidelity of the SBS process [7.6], as described previously.

7.5   Laser beam combining using SBS

In the normal SBS configuration a single pump beam is incident upon the SBS medium, and the Stokes scattered output beam is generated by amplifi-cation of noise. This SBS scattered field will have a random overall phase since it starts from statistical noise and, in addition, this phase has been shown to fluctuate randomly in a time of the order of several times the phonon lifetime [7.14]. The SBS reflection, therefore, has no absolute temporal phase reference. As a consequence, if two beams are conjugated by SBS in separate interaction volumes the two Stokes beams will have a phase difference that is random and unrelated to the phase difference of the pump beams. Basov *et al* [7.15] first predicted this random phase differ-ence and several groups demonstrated its existence [7.16–7.18]. If the random phase difference in the Stokes beams is to be avoided an absolute phase reference must be created. This can be accomplished by several methods.

7.5.1   Laser beam combining using spatial overlap in SBS

Consider the case where two or more pump beams are overlapped in the same interaction volume. They appear to the SBS process as a single, but highly aberrated, beam with the relative phase between the beams appearing as an aberrated wave front. The standard mechanisms that phase conjugate a single aberrated pump beam will be responsible for phase conjugation of the entire ensemble of interacting beams and such that the phases of the Stokes beams are locked in relation to the original pump beam phases.

Basov [7.19] has achieved phase locking by overlapping two aberrated beams in a light guide within a Brillouin cell, so that complete overlap is achieved in the SBS interaction region. Another method that has been widely used involves the overlapping of a number of beams at the focal point of a lens [7.20–7.22]. However, the overlap in this case can be very sensitive to small changes in the angles between the beams [7.23].

7.5.2 Laser beam combining using back-injection of a Stokes seed

The phases of the two Stokes beams may be locked, even though the beams are conjugated in separate interaction volumes, if the initiating Stokes beams are generated by amplification of a common Stokes seed beam [7.24–7.26]. This Stokes beam can be produced by SBS reflection in a separate SBS cell. Back-injection of a Stokes seed beam also permits the reduction of the SBS threshold and an increase in reflectivity. Some limitation on beam-combination efficiency due to the statistical nature of SBS are reported in [7.25].

7.5.3 Laser beam combining using BEFWM

Brillouin enhanced four wave mixing (BEFWM) has already been described in chapter 6. BEFWM has been used to phase lock beams [7.27–7.29]. In the geometry depicted in figure 7.18, a counter-propagating Stokes wave E_2 that is the phase conjugate of a pump beam E_1 [7.29] together act as a pair of phase conjugate pump waves. The counter-propagating Stokes pump E_2 can be generated by an SBS loop geometry from the beam E_1 [7.29], as this configuration suffers less from random phase jumps.

If a beam E_3 (at the same pump frequency as beam E_1) overlaps with beams E_1 and E_2, a beam that is phase conjugate to E_3 can be produced by FWM. In this configuration, beam E_3 interacts with E_2 to write an acoustic grating in the Brillouin medium. When E_1 scatters from this grating it produces a beam that is phase conjugate to E_3 and phase-locked to E_2.

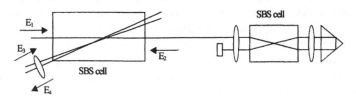

Figure 7.18. A schematic diagram of the experimental configuration for phase-locked phase conjugation using Brillouin enhanced four-wave mixing.

Since E_2 is also the phase conjugate of E_1 the two conjugates E_2 and E_4 of two input beams are phase locked. Any additional beams that are overlapped at any point within the volume illuminated by E_1 and E_2 are also phase conjugated with their phases locked to the phase of E_2. Unlike in the usual overlapping-focus SBS scheme, there is no requirement for the different focused beams to overlap mutually. Thus there is much less sensitivity to changes in alignment.

The requirement for good phase locking is for pumps to be much stronger than the spontaneous scattering noise: $I_1 I_2 \gg I_3^2 \exp(-25)$. In experiments, correlation between poor phase locking and phase fluctuations was observed, suggesting that the fluctuations in the outputs might be occurring not because the FWM interaction was too weak but rather because the fluctuations in the phase of E_2 were disrupting the phase-locking process. The use of SBS with optical feedback using a loop scheme to produce E_2 has been used to suppress phase fluctuations in the SBS output [7.29].

7.6   Laser pulse compression by backward SBS

One important phenomenon which accompanies SBS is laser pulse compression. The basic aim of pulse compression by backward wave amplification involves taking an optical pulse of relatively long duration and converting it to a shorter duration without serious loss of energy [7.30]. As a result, a more powerful pulse is created than the original input and the technique is therefore quite distinct from merely 'chopping out' a short duration pulse from a longer one by electro-optic techniques or by using a saturable absorber, for example.

A simple geometrical form of the pulse compression scheme is illustrated in figure 7.19. The long laser pulse (frequency ω_L) of duration t_L enters the front end of a Brillouin-active (or Raman-active) medium, whose length is approximately half that of the laser pulse ($L \approx ct_L/2$). A Stokes pulse (frequency ω_S) of short duration is arranged to propagate from the back end of the interaction zone just as the leading edge of the laser pump pulse arrives. As it travels in the backward direction, energy is coupled from the laser pulse into the Stokes pulse via the material resonance ($\omega_B = \omega_L - \omega_S$). Consequently, the backward Stokes pulse is amplified and the pump laser is depleted when the Stokes pulse has grown to a saturating fluence. By using the optimal interaction length the Stokes pulse emerges just as the end of the laser pulse has entered the medium. An intense pulse of short duration is created and if the input pulse is strongly depleted, the output pulse will be more powerful than the input. Optical pulse compression by backward wave amplification is achieved.

The two important parameters of the interaction are the compression ratio ($K = t_L/t_S$) defined as the ratio of the duration of the input laser

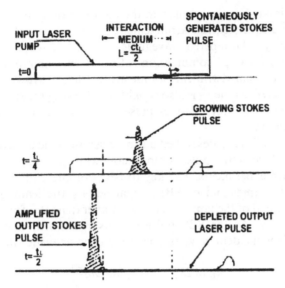

INPUT LASER
PUMP

INTERACTION
↦— MEDIUM ··· ⋈
$L = \frac{ct_L}{2}$

SPONTANEOUSLY
GENERATED STOKES
PULSE

t=0

GROWING STOKES
PULSE

$t = \frac{t_L}{4}$

AMPLIFIED
OUTPUT STOKES
PULSE

$t = \frac{t_L}{2}$

DEPLETED OUTPUT
LASER PULSE

Figure 7.19. Basic scheme of pulse compression by stimulated Brillouin backward scattering.

pulse (t_L) to the duration of the output Stokes pulse (t_S), and the energy conversion efficiency ($\eta = E_S^{OUT}/E_L^{IN}$) of the interaction. The enhancement in power is also given approximately by the product of these two quantities $G_p \approx K\eta$. The energy conversion efficiency is a well defined quantity that can readily be measured. It is also at times convenient to talk in terms of the photon conversion efficiency (η_p), especially in numerical and analytical modelling of the interaction, since it is the total number of photons that is normally conserved if optical losses are not significant. The two quantities of conversion efficiency are simply related as $\eta = (\omega_S/\omega_L)\eta_p$. In SRS the quantum efficiency (ω_S/ω_L) needs to be considered due to an appreciable frequency shift (($\omega_L - \omega_S)/\omega_L \approx 8\%$ with KrF 248 nm laser in methane). In Brillouin scattering the frequency shift is so small ($\omega_B/\omega_L \approx 10^{-5}$) that the photon conversion efficiency can be taken as identical to energy conversion efficiency for all practical purposes.

Unlike the conversion efficiency, the pulse compression ratio is not as rigorously well defined. The Stokes pulse shape can be totally different from the input pulse shape and the usual definition of pulse duration as full width at half maximum (FWHM) height can in some cases be misleading. In some cases it also is important to know detailed features of the Stokes pulse, such as the rise-time of its leading edge (not implicit in the definition of the compression ratio), e.g. to provide a knowledge of the pre-pulse heating of a target in laser–plasma studies that the wings on the leading edge of the pulse can produce.

The collimated geometry used in the compressor systems, although the simplest case and the easiest arrangement to model, is not the only interaction geometry. The tapered waveguide is in many ways superior to the collimated case for large compression ratios. It will maintain a low Stokes intensity and prevent detriment nonlinearities that a growing Stokes may cause. The tapered geometry is also capable of allowing the spontaneous generation of a short duration Stokes pulse by using a tapered waveguide or focusing geometry.

Hon [7.31, 7.32] has developed a simple semi-classical theory that offers an intuitive picture of SBS pulse compression, while providing results in good agreement with experiments. The key to SBS compression was the tapered light guide. The threshold of SBS is reached by the leading edge of the pulse at the far end of the taper where the smaller diameter forces an increase of power density. As the SBS pulse sweeps backward, it beats with the remainder of the incident wave to create a strong acoustic wave with

$$k_B \approx 2k_L, \qquad \omega_B \approx 2nv\frac{\omega_L}{c} \qquad (7.10)$$

which in turn acts as a bulk grating to reflect the incident wave further to strengthen the SBS wave coherently.

It was postulated that the leading edge of this phonon envelope forms the mirror that reflects and, owing to its growing reflectivity, compresses the pulse.

The reflectivity, r, of an acoustic wave, when the Bragg condition is satisfied, as is the case here, is given by

$$r \approx \sin^2(\pi l\sqrt{M_2 I_0}/2\lambda), \qquad (7.11)$$

where l is the interaction depth, I_0 is the acoustic power density and M_2 is the acousto-optic figure of merit.

The propagation of a square incident pulse is assumed in a uniform light guide of diameter d and the presence of an interaction region with length L, in which SBS occurs. The acoustic power density is

$$I_0 \approx E_0 v_0/Ld^2 \qquad (7.12)$$

where E_0 is the acoustic energy produced within the volume Ld^2. Setting the SBS efficiency $r = 1$, one can obtain the penetration depth of the acoustic wave

$$l \approx (\lambda c/2n^4 p_{12})\sqrt{\rho v_0/P} \qquad (7.13)$$

where P is the power density of the incident light and p_{12} is the elasto-optic constant. The compressed pulse width is therefore simply

$$\delta T \approx nl/c \approx (\lambda/2n^3 p_{12})\sqrt{\rho v_0/P}. \qquad (7.14)$$

Using equation (7.14) and considering pressurized methane as the non-linear medium, the calculated compressed pulse, $\delta T = 2.1$ ns, agrees very well with the observed values of 2 ± 0.5 ns.

Damzen *et al* [7.33] have numerically modelled compression in a wave-guide with a convergent taper. The dynamics of the process is investigated in this geometry and particular attention is paid to the parameters which control the compression ratio and the conversion efficiency of the process. The authors showed that, as the laser propagates along the waveguide into a decreasing area, its intensity increases. The gain is greatest near the exit of the waveguide and, by suitable choice of input laser intensity and waveguide taper, a Stokes pulse of short duration originates from a region near the exit. Once generated, it propagates backwards down the waveguide, receiving amplification by the incoming laser pulse. An intense Stokes pulse of short duration is produced and the laser pulse is correspondingly depleted.

The Stokes field, e_S, the laser field, e_L, and the acoustic fluctuation, q, were represented in the form

$$e_L(z,t) = \tfrac{1}{2}E_L(z,t)\exp \mathrm{i}(\omega_L t - k_L z) + \text{c.c.}$$

$$e_S(z,t) = \tfrac{1}{2}E_S(z,t)\exp \mathrm{i}(\omega_S t - k_S z) + \text{c.c.} \tag{7.15}$$

$$q(z,t) = \tfrac{1}{2}Q(z,t)\exp \mathrm{i}(\omega_q t + k_q z) + \text{c.c.}$$

Using the slowly varying envelope approximation, the resulting coupled wave equations are taken into the following form

$$\left(\frac{\partial}{\partial t} + \frac{c}{n}\frac{\partial}{\partial z}\right)E_L = -E_S Q$$

$$\left(\frac{\partial}{\partial t} - \frac{c}{n}\frac{\partial}{\partial z}\right)E_S = E_L Q^* \tag{7.16}$$

$$\left(\frac{\partial}{\partial t} + \frac{1}{2\tau_B}\right)Q = \gamma E_L E_L^*$$

where τ_B is the damping time of the acoustic wave in the Brillouin-active medium and γ is a coupling constant which is proportional to g_B/τ_B, where g_B is the steady-state gain coefficient.

These equations are appropriate to the transient regime when the second derivatives, the higher order scattering and the optical losses can be neglected.

A fourth-order Runge–Kutta scheme was used to solve numerically the system of equations. The tapered geometry required for pulse compression was modelled by making the gain term γ a function of z related to the waveguide area $A(z)$ in the form $\gamma(z) = \gamma_0/A(z)$.

Both squared input laser pulses, $I_L = I_{L0}$ and Gaussian input pulses, $I_L = I_{L0}\exp -(\tau/\tau_0)^2$, are considered. The initial Stokes field is of the form: $I_L = I_L\,\mathrm{e}^{-G}$, where $G \approx 30$ under many experimental conditions.

Computations have considered the Brillouin-active methane at high pressure as a representative medium exhibiting high gain, g_B, and long damping time, τ_B. The pulse output contains features of compression with a Stokes pulse of shorter duration than the input laser pulse and correspondingly depleted output laser pulse. However, the output Stokes pulse is of a complex nature exhibiting pulse break-up under transient SBS conditions. Break-up of the Stokes pulse would be mostly eliminated if the Brillouin-active medium were strongly damped, since the laser field would be regenerated from a rapidly diminishing acoustic field. Ideally, the damping time should be less than the duration of the Stokes pulse. The limit of compression by this technique appears to be reached when the Stokes pulse is reduced to a duration comparable with the inverse of the acoustic frequency.

The main disadvantages of SBS as a compressor technique lie in its transient nature. The laser pump must have narrow linewidth for coherent interaction of the laser and Stokes fields. The effects of Stokes pulse break-up must be accounted for under transient conditions and can be controlled by a sufficiently large taper on the waveguide.

Highly efficient compression of laser pulses by stimulated Brillouin scattering in carbon disulphide and highly-pressurized methane down to 1 ns in duration, has been demonstrated by Damzen *et al.* Compression ratios of ~10 and energy conversion efficiencies >70% have been produced. Several compressor systems have been investigated including the use of tapered waveguides, long focal length geometries and generator–amplifier systems [7.34].

The SBS and SRS processes were used also in order to produce UV laser beams of short time duration and good optical quality [7.35, 7.36]. The oscillator pulse was 11 ns long while the amplified phase conjugate beam duration could vary from 3.3 to 1.5 ns due to the compression by SBS. When this last laser beam was focused into a Raman cell containing methane at 30 atm. the shortest backward stimulated Raman scattering pulse was 170 ps.

7.7 Conclusions

SBS with optical feedback has been experimentally and theoretically demonstrated to reduce significantly the threshold power and enhance reflectivity. Transient effects are shown to be important and the various parameters, which control the performance of such feedback devices in the transient regime, have been investigated numerically. The numerical simulations are in good agreement with the experimental results.

The results presented have shown that a two-cell generator–amplifier system is an effective way of obtaining phase conjugation of high-power radiation by the SBS process. Numerical simulations have been presented and shown to provide a reasonable prediction of the transient response of the experimental system. Similarly the steady-state calculations are in agreement with the broad numerical and experimental trends. This technique could be used to scale to even higher laser powers by introducing further cells with progressively decreasing apertures. An alternative would be to use a tapering amplifier cell.

Several schemes also exist to provide the conditions under which SBS can be used to coherently combine beams.

Backward SBS in long interaction length cells can produce a considerable degree of pulse compression with high efficiency with the limit of compression down towards the inverse of the acoustic frequency.

References

7.1 Odintsov V I and Rogacheva L F 1983 *JETP Lett.* **36** 344
7.2 Wong G K N and Damzen M J 1988 *J. Mod Opt.* **35** 483
7.3 Scott A M and Hazell M S 1986 *IEEE J. Quantum Electron.* **QE-22** 1248
7.4 Basov N B, Zubarev I G, Kotov A V, Mirinov A B, Mikhailov S I, Pasmanik G A, Smirnov M G and Shilov A A 1979 *Sov. J. Quantum Electron.* **9** 237
7.5 Efimkov V F, Zubarev I G, Kotov A V, Mirinov A B, Mikhailov S I, Pasmanik G A, Smirnov M G and Shilov A A 1979 *Sov. Phys. JEPT* **50** 267
7.6 Wong G K N and Damzen M J 1990 IEEE *J. Quantum Electron.* **26** 139
7.7 Zeldovich B Ya, Pilipetsky N F and Shkunov V V 1985 *Principles of Phase Conjugation* vol 42, Springer Series in Optical Science (New York: Springer) ch 4
7.8 Hull D R, Lamb R A and Digman J R 1989 *Opt. Commun.* **72** 104
7.9 Glazkov D A, Efimkov V F, Zubarev I G, Pastukhov S A and Sobolev V B 1988 *Sov. J. Quantum Electron.* **18** 974
7.10 Varlamova I A, Golubev V V and Sirazetdinov V S 1989 *Sov. J. Quantum Electron.* **19** 1631
7.11 Crofts G J and Damzen M J 1991 *Opt. Commun.* **81** 237
7.12 Bubis E L, Konchalina L R and Shilov A A 1989 *Sov. J. Quantum Electron.* **19** 925
7.13 Crofts G J and Damzen M J 1991 *J. Opt. Soc. Am. B* **8** 2282
7.14 Mangir M S, Ottusch J J, Jones D C and Rockwell D A 1992 *Phys. Rev. Lett.* **68** 1702
7.15 Basov N G, Efimkov V F, Kotov A V, Mironov A B, Mikhailov S I and Smirnov M G 1981 *Sov. J. Quantum Electron.* **11** 1335
7.16 Bespalov V I, Betin A A, Pasmanik G A and Shilov A A 1980 *JETP Lett.* **31** 630
7.17 Vasil'ev M V, Gyulameryan A L, Mamaev A V, Ragul'skii V V, Semenov P M and Sidorovich V G 1980 *JETP Lett.* **31** 634
7.18 Basov N G, Zubarev I G, Mironov A B, Mikhailov S I and Okulov A Y 1980 *Sov. Phys. JETP* **52** 847
7.19 Basov N G, Efimkov V F, Zubarev I G, Kotov A V, Mikhailov S I and Smirnov M G 1978 *JETP Lett.* **28** 197
7.20 Basov N G, Efimkov V F, Zubarev I G, Kotov A V, Mironov A B, Mikhailov S I and Smirnov M G 1979 *Sov. J. Quantum Electron.* **9** 455
7.21 Rockwell D A and Giuliano C R 1986 *Opt. Lett.* **11** 147
7.22 Valley M, Lombardi G and Aprahamian R 1986 *J. Opt. Soc. Am. B* **3** 1492
7.23 Sternklar S, Chomsky D, Jackel S and Zigler A 1990 *Opt. Lett.* **15** 469

7.25 Falk J, Kanefski M and Suni P 1988 *Opt. Lett.* **13** 39
7.25 Caroll D L, Johnson R, Pfeifer S J and Moyer R H 1992 *J. Opt. Soc. Am. B* **9** 2214
7.26 Andreev N F, Khazanov E A, Kuznetsov S V, Pasmanik G A, Shklovsky E I and
        Sidorin V S 1991 *IEEE J. Quantum Electron.* **27** 135
7.27 Ridley K D 1995 *J. Opt. Soc. Am. B* **12** 1924
7.27 Ridley K D and Scott A M 1996 *J. Opt. Soc. Am. B* **13** 900
7.29 Scott A M, Whitney W T and Duignan M T 1994 *J. Opt. Soc. Am. B* **11** 2079
7.30 Damzen M J 1983 PhD Thesis, The Blackett Lab., IC, London
7.31 Hon D T 1980 *Opt. Lett.* **5** 516
7.32 Hon D T 1982 *Opt. Eng.* **21** 252
7.33 Damzen M J and Hutchinson M H R 1983 *IEEE J. Quantum Electron.* **QE-19** 7
7.34 Damzen M J and Hutchinson M H R 1983 *Opt. Lett.* **6** 313
7.35 Filippo A and Perrone M R *1992 J. Mod Optics* **39** 1829
7.36 Nassisi V and Pecoraro A 1993 *IEEE J. Quantum Electron.* **29** 2547

Chapter 8

SBS in optical fibres

The power needed to generate SBS in bulk (non-guiding) materials is of the order of 10^5 W. This power can be reduced by increasing the interaction length and by decreasing the cross-sectional area of the light beam. These requirements are satisfied by waveguides, among which the optical fibre is the most attractive.

SBS in optical fibres is relatively easy to realize. Since 1972, when SBS was observed for the first time, in single-mode fibres, by Ippen and Stolen [8.1], there have been many investigations of this topic [8.2–8.5]. The issues associated with SBS in optical fibres are of significant interest in optical communications. One issue is the limitation of power transmission along fibres with narrowband radiation. However, there are possible applications of SBS in fibres. This subject will be treated in the latter section of this chapter.

It should be noted that most research was concentrated principally on single-mode fibres, which are not suitable for optical phase conjugation (OPC). In multimode fibres, phase conjugation by SBS was obtained for the first time, by Petrov and Kuzin and Basiev *et al*, in 1982 [8.5]. OPC can only be achieved in multimode fibres, because only these fibres allow input and propagation of an aberrated beam quality.

8.1 Phase conjugation by SBS in optical fibres

The first OPC experiments in multimode fibres were carried out with pump powers in the range 1–10 kW, with pulse duration of several tens of nano-seconds. The threshold powers of SBS generation were found to be: 1 kW for pumping by the second harmonic and 4 kW for pumping by the first harmonic of the YAG laser. To ascertain that OPC indeed took place, the spatial distribution pattern of the Stokes wave field was photographed and compared with the spatial distribution of the pump field, and a good similarity was obtained. The pump to Stokes energy conversion efficiency was about 50% with a multimode fibre of core diameter 30 μm and numerical aperture 0.15. More recently, OPC by SBS in optical fibres was achieved at low pump powers ∼10 W [8.5].

Recent experiments with multimode fibres were made by Eichler *et al* [8.6], using glass fibres (undoped multimode quartz fibres). The optical fibres were of a core diameter of 200 µm and different lengths. Using a Nd:YAG laser with a 50 cm coherence length and near diffracted limited and a fibre of 17 m length, the SBS threshold was found to be 17 kW, the SBS reflectivity up to 50% and SBS fidelity up to 93%. The influence of the coherence length of the input beams was found to be a decisive factor for the limitation of the SBS interaction length in the fibre. The conclusion was that, if the linewidth of the single laser mode becomes small compared with the Brillouin linewidth, the interaction length of SBS in the fibre is no longer limited by the coherence of the pump beam but only by fibre attenuation.

Fibre phase conjugators developed are harmless to the environment and can be handled very easily, in contrast to the liquid and gaseous SBS media. The quartz multimode fibre system has a dynamic range as high as 20 times the threshold energy. The excellent parameters of novel fibre phase conjugators in standard or tapered geometry beat the performance of all liquid or gas SBS cells, used up to now, in many aspects.

Similar data were achieved in the visible range at a wavelength of 532 nm and in the ultraviolet at 355 nm [8.7].

SBS can be detrimental for optical communication systems (see section 8.3.1). At the same time it can be useful as a method of improving beam quality via phase conjugation of high-power solid-state lasers used in many application fields in industry and science [8.8–8.11]. Power scaling of solid-state lasers under conservation of their beam profiles is possible using novel fibre phase conjugators [8.12].

Heuer [8.13, 8.14] has demonstrated SBS in a novel optical fibre phase conjugated mirror as a generator–amplifier system using a tapered quartz fibre. In this case the threshold energy was reduced to 15 µJ and the SBS reflectivity was 92%. The fibre could be operated to a dynamic power range of 1:267. The phase conjugation fidelity was measured to be greater than 95%, over the entire pump energy range.

Due to the low SBS threshold (power below 1 kW) fibre phase conjugators are suitable for cw high power lasers developed by Harrison *et al* [8.15] and Kovalev and Harrison [8.16] and quasi-cw high power lasers developed by Eichler *et al* [8.17, 8.18].

SBS in long multimode optical fibres has been demonstrated by Rodgers *et al* [8.19] as a new technique for coherent combining of low-power laser beams by use of semiconductor diode lasers. Very recently a theoretical model for diffraction-limited high-power multimode fibre amplifiers using seeded SBS phase conjugation (SBS beam clean up) have been published [8.20].

8.2   Theoretical model of phase conjugation by SBS in optical fibres

Following Kuzin *et al* [8.5] and Hellwarth [8.21], one can develop a theoretical model to explain the OPC by SBS in optical fibres. For this analysis we assume the pump field amplitude is not depleted during the SBS process. The Stokes field evolution is given, as before, by

$$\nabla^2 \mathbf{E_S} - \frac{n^2}{c^2}\frac{\partial^2}{\partial t^2}\mathbf{E_S} = \mu_0 \frac{\partial^2}{\partial t^2}\mathbf{P}^{NL}. \tag{8.1}$$

For convenience we define a nonlinear polarization expressed by

$$\mathbf{P}^{NL} \cong i\varepsilon_0 \chi^{(3)}(\mathbf{E_p}\mathbf{E_p})\mathbf{E_S} \tag{8.2}$$

where $\chi^{(3)}$ is the imaginary part of the nonlinear susceptibility of the medium and E_p is the pump field amplitude. In the steady-state, equation (8.1) takes the form

$$\frac{d^2}{d\varphi_S^2}(E_S^* \, e^{-i\varphi_S}) + (\varepsilon + i\chi^{(3)}|E_P^*|^2)(E_S^* \, e^{-i\varphi_S}) = 0 \tag{8.3}$$

where E_S^* is the complex conjugate of the Stokes field and φ_S is its phase (kz) where k is the optical wavevector. Introducing in equation (8.3) the slowly varying envelope approximation (SVEA)

$$\left|\frac{d^2 E_S^*}{dz^2}\right| \ll \left|K_S \frac{dE_S^*}{dz}\right| \tag{8.4}$$

one can obtain the evolution equation for the Stokes field

$$\frac{dE_S^*(\varphi_S)}{d\varphi_S} = \frac{1}{2\varepsilon}\chi^{(3)}|E_P^*|^2 E_S^* \tag{8.5}$$

and

$$E_S^*(\varphi_S) = E_S^*(0)\exp[(k_S/2\varepsilon)\chi^{(3)}|E_P^*|^2 \varphi_S]. \tag{8.6}$$

In terms of intensities and propagation variable (z), one can find

$$I_S(z) = I_S(0)\exp(g_B I_P z) \tag{8.7}$$

where $g_B = 5 \times 10^{-9}$ cm/W for quartz fibres and $I_S(0)$ should be taken as the intensity of the spontaneous Brillouin scattering. The SBS threshold is usually defined for $\exp(g_B I_p L) \approx 20$, in the fibre case.

For multimode fibres, the pump and Stokes fields have to be decomposed on the system of fibre modes

$$E_P^* = \sum_{(l)} E_{P,l}^*(\varphi_S) f_l(x',y') e^{i\beta_l \varphi_S}$$

$$E_S^* = \sum_{(m)} E_{S,m}^*(\varphi_S) f_m(x',y') e^{-i\beta_m \varphi_S}$$

(8.8)

where $f_l(x',y')$ are orthonormal functions describing the spatial distribution of the fibre modes and β_l are the mode propagation constants. Introducing equations (8.8) into the evolution equation (8.1) and using again SVEA, one gets the Stokes mode amplitudes

$$\sum_{(m)} \{\beta_m f_m(x',y')\} e^{-i\beta_m \varphi_S} \frac{d\bar{E}_{Sm}(\varphi_S)}{d\varphi_S} - \frac{1}{2\varepsilon} \chi^{(3)} \bar{E}_{Sm} |\bar{E}_p|^2 f_m(x',y') e^{-i\beta_m \varphi_S} = 0.$$

(8.9)

Multiplying equation (8.9) by f_m^* and integrating over the fibre cross-section, one can obtain the equation system describing the evolution of the Stokes field on the fibre modes:

$$\frac{dE_{Sl}^*(\varphi_S)}{d\varphi_S} = \frac{L}{2\varepsilon\beta_l} \chi^{(3)} \sum_{m=1}^{N} E_{Sm}^*(\varphi_S) e^{-i(\beta_m - \beta_l)\varphi_S}$$

$$\times \iint_{(S)} |E_P|^2 f_m(x',y') f_l^*(x',y') \, dx' \, dy', \qquad l = 0,1,\ldots,N \quad (8.10)$$

where N is the number of modes in the fibre and S is the fibre cross-sectional area. These equations and correct phases between the modes allow the phase conjugation of the Stokes wave with respect to the pump one.

Hellwarth [8.21] was looking for the solution of the Stokes mode equations, which ensures the same gain coefficient, γ_F, for all modes:

$$E_n^*(\varphi_S) = C_n e^{\gamma_F \cdot \varphi_S/2}.$$

(8.11)

In this case, equation (8.11) leads to the eigenvalue system

$$\frac{\gamma_F c^2}{4\pi\omega_S^2} K_S^2 C_n = \sum_{(m)} D_{nm} C_m \Bigg|_{n=1,2,\ldots,N}$$

(8.12)

where

$$D_{m,n} = \sum_{(i,j)} \iint_{(S_F)} dx' \, dy' f_m^* f_i f_j^* f_n K_{mijn} \mathbf{E}_{P_i} \cdot \mathbf{E}_{P_j} / E_0^2$$

(8.13)

$$K_{mijn} = \frac{1}{K_S L} \int e^{i\Delta\beta\varphi_S} \, d\varphi_S$$

(8.14)

and
$$\Delta\beta = \beta_m^S + \beta_i^P - \beta_j^P - \beta_n^S. \tag{8.15}$$

The main contribution to the coefficients D_{mn} is brought by the terms with $\Delta\beta = 0$ and $K_{mijn} = 1$, i.e. the terms with $m = n$, $m = j$ and $i = n$. All other terms have $\Delta\beta$ very different from zero and $K_{mijn} \ll 1$.

The further analysis of the eigenvalue problem shows that the Stokes wave, which is the phase conjugate of the pump one $(C_n = E_{P_n}^*)$, will dominate the process and this process is stronger as the number of modes is higher. It has been shown that the frequency shift of the Stokes wave leads to an imperfect optical phase conjugation (OPC), which is an undesirable effect increasing with the fibre length. If one accepts a non-OPC fraction, r_{nOPC}, the fibre length, ensuring this fraction, is

$$L_{OPC} \le 6r_{nOPC}^{1/2} S/N\Delta\lambda. \tag{8.16}$$

For usual fibres, if $r_{nOPC} = 0.1$, the fibre length should not be greater than several metres.

Other numerical models of phase conjugation by SBS can be found in [8.22–8.25].

8.3 Experiments and results in phase conjugation by SBS in optical fibres

An example of a system used to study the reflectivity of fibres is shown in figure 8.1 [8.26]. The laser system was an oscillator–amplifier based on Nd:YAG flash lamp pumped, with Q-switched oscillator operating TEM$_{00}$ and narrow-linewidth. An optical isolator, composed of two Glan polarizers and a Faraday rotator, was used to decouple laser from SBS reflection. A half-wave plate was used to control the energy from the laser system entering the fibre.

In order to measure the reflectivity of the fibre, a partially reflecting mirror was placed into the beam to get a reference for the pumping beam and the back-reflected Stokes beam measured on an energy metre (EM). For coupling the beam into the fibres, antireflection-coated plane convex lenses with different focal lengths were used. An He–Ne laser was used to provide a visible reference for coupling adjustment. It was not possible to couple all the pumping energy into the fibre. Using the energy meter (EM2) a measurement of the transmitted energy, ε_{trans}, through the fibre was made.

Taking into account that 4% of energy is lost at the exit of the fibre, the formula for coupling efficiency is:

$$C_{eff} = \frac{\varepsilon_{trans}/0.96}{\varepsilon_{pump}}. \tag{8.17}$$

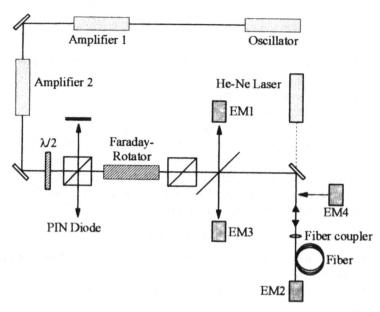

Figure 8.1. Setup for study of the phase conjugation by SBS in multimode optical fibres.

The maximal coupling efficiency for the fibres investigated was measured to be 85% and up to 90% in some cases.

The SBS internal reflectivity of the fibre is the ratio between the Stokes
energy and the energy coupled into the fibre, ε_{in}. Taking into account again
the Fresnel losses, the formula for the reflectivity is

$$R_{SBS} = \frac{\varepsilon_S}{\varepsilon_{in}} = \frac{\varepsilon_{refl}/0.96}{C_{eff}\varepsilon_{pump}} \tag{8.18}$$

where ε_{refl} is the Stokes energy measured outside the fibre, by energy meter
(EM3).

Multimode quartz glass fibres with index step stucture and with different
core diameters have been used to study the SBS threshold, damage threshold
and SBS reflectivity. The dependence of the SBS reflectivity versus pumping
energy for a core diameter of 25 μm is illustrated in figure 8.2. The behaviour
of the fibres with core diameters of 50, 100 and 200 μm is similar.

A comparison of the dependence of the experimental fibre reflectivities
against the pulse peak power is shown in figure 8.3.

For all fibres the reflectivity's curves were plotted until destruction of the
fibre entrance surface occurred. The corresponding energy was used to
calculate the damage threshold. The damage threshold of the fibre entrance
surface lies above $0.9\,\text{GW/cm}^2$ if homogeneous illumination of the entire
core diameter of the fibre by laser radiation is assumed. The results of the
measurements are summarized in table 8.1. Measurement of stimulated

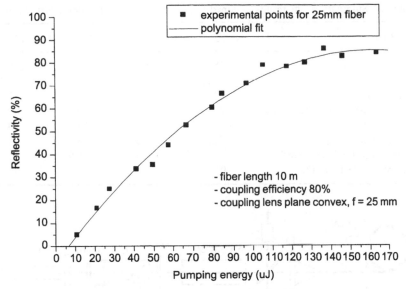

Figure 8.2. SBS reflectivity against pumping energy for the 25 µm.

Brillouin scattering threshold for various types of fibre can be found also in [8.27].

A system to study the temporal behaviour of SBS in multimode fibres [8.28] is shown in figure 8.4. The laser is an Nd:YAG oscillator–amplifier operated in the Q-switched mode and with reduced spectral linewidth (the coherence length is 30 cm). The duration of the laser pulse is 18 ns and the

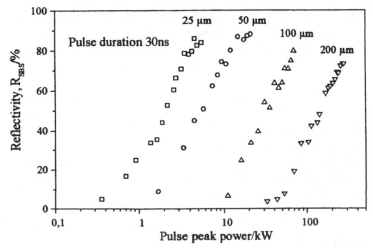

Figure 8.3. Comparison of the SBS reflectivity of the fibres investigated against pulse peak power.

Table 8.1. Summary of the results obtained regarding SBS in multimode quartz fibres.

Core (μm)	25	50	100	200
Clad (μm)	125	125	125	220
NA	0.22	0.22	0.22	0.22
Length (m)	10	10	10	7
Coupling lens focal length (mm)	25	40	80	100
Coupling efficiency (%)	80	80	85	80
SBS threshold (kW)	0.3	1.6	8.3	26
Damage threshold (GW/cm^2)	1.0	1.0	0.9	1.0
Maximum reflectivity (%)	90	88	88	80

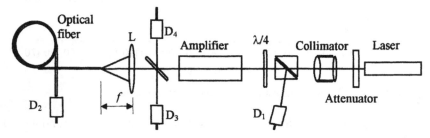

Figure 8.4. Setup for studying the temporal behaviour of SBS in multimode optical fibres.

maximum output energy was 4 mJ in a beam with a near Gaussian transversal distribution. The beam was focused with a lens with short focal length (16 cm) in a quartz fibre with 200 μm core diameter, 2 m length and 0.2 numerical aperture, which had a polished entrance surface.

Time measurements with a fast photodiode (rise time of 1 ns) and a digital oscilloscope have been performed. Figure 8.5(a) shows the oscilloscope trace of the incident laser of 18 ns duration. Figures 8.5(b) and (c) show the transmitted pump and the Stokes pulse, respectively, near the threshold.

One can observe that the Stokes pulse is two times shorter, so it is possible to consider that SBS takes place in transient conditions. Under increased pumping power, the transmitted and the Stokes signals exhibit multiple pulse behaviour (figures 8.6(a) and (b)).

This behaviour is specific for the transient regime and was also observed in [8.29], in the tapered geometry with liquid SBS media. The particular time dependence of the Stokes and transmitted pulse can be explained by the interaction between the Stokes pulse and the incident laser pulse, which may exchange energy back and forth during the pumping pulse duration. The Stokes pulse is generated by the interaction between the incident laser field and the acoustic field, so when the Stokes pulse is maximum, the transmitted pulse is minimum. When the Stokes pulse propagates through the fibre, the

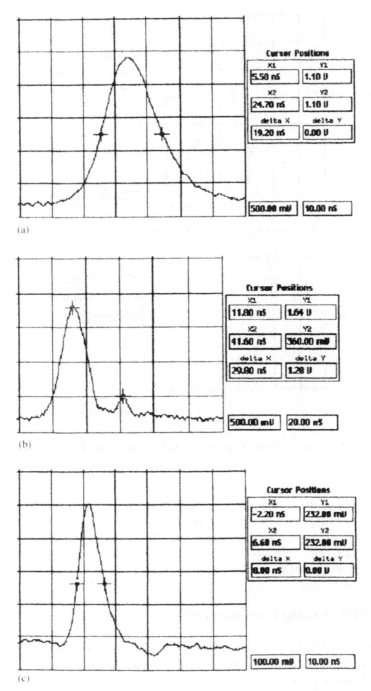

Figure 8.5. The temporal form of the incident (a), transmitted, (b) and Stokes (c) pulses in SBS in an optical fibre.

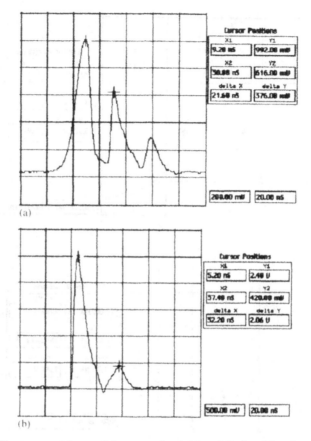

Figure 8.6. The temporal form of the transmitted (a) and Stokes (b) pulses in SBS in an optical fibre.

leading edge of the Stokes pulse increases in intensity but its tail can continuously lose energy back to the laser pulse. The regenerated laser pulse has a phase shift of π and can generate a new Stokes pulse in interaction with the acoustic field, if it has enough energy. The process is repeating, until the energy decreases under the SBS threshold.

8.4   SBS in optical communications

Wide-bandwidth, long-distance data communication systems are based on coherent optical transmission in single-mode fibres using narrowband single-frequency lasers.

SBS is one of the most dominant nonlinear effects in single-mode optical fibres. It is particularly important when narrow-linewidth laser light is launched into long single-mode fibres. The narrow-linewidth excitation, the long interaction length and the high power densities contribute to low-threshold SBS. The Brillouin threshold can be quite low (~1 mW) for 1.55 μm optical communication systems employing low-loss optical fibres ($\alpha = 0.2$ dB/km).

8.4.1 Harmful SBS effects in optical communication systems

It is well known that SBS could limit the information transfer through optical fibres by limiting the maximum transmitted power, reflecting power back into the light source and generating extraneous channels having shifted frequencies [8.30].

SBS can be detrimental in a number of ways. First, as a result of the power transfer of the forward-travelling pump light to the backward-travelling Stokes wave, a much lower signal than expected in the absence of SBS will be delivered to the receiver. Second, the Stokes wave provides a backward coupling into the transmitter that can destabilize the laser operation [8.31]. Third, it is possible for the backward SBS pulse to attain a peak power greatly exceeding that of the input wave. When the power in such a pulse becomes great enough permanent damage to the fibre can result [8.1].

There is also another effect that may be caused by SBS in single-mode fibres. If SBS occurs in the presence of external feedback the transmitted intensity as well as the SBS intensity will oscillate [8.32]. The bandwidth of the noise presented by this mechanism has been measured to cover a range of 10–20 MHz at $\lambda = 514.5$ nm. However, this bandwidth of the noise would be narrower for longer wavelengths, since the Brillouin-gain width becomes narrower. It is emphasized that steady-state oscillations occur even at very low reflectivities, which may be caused by the cleaved ends of the fibre or other sources in a communication system and therefore must be taken into account.

For the reasons mentioned above it is important to keep the transmitted power injected into the fibre below the SBS threshold.

One way to increase the input power launched into the fibre is to raise the SBS threshold. For cw or quasi-cw pumps, the Brillouin-gain is substantially reduced if the pump spectral width ($\Delta\nu_p$) exceeds the Brillouin bandwidth ($\Delta\nu_B$) [8.2]. Detailed calculations [8.33] show that the Brillouin-gain under broadband pumping conditions depends on the relative magnitudes of the pump coherence length (L_{coh}) and the SBS interaction length (L_{int}). The Brillouin-gain is significantly reduced if $L_{coh} \ll L_{int}$. This situation is generally applicable to optical fibres where the interaction length is of the order of the fibre length. The Brillouin-gain is reduced by a factor $\Delta\nu_p/\Delta\nu_B$ for $\Delta\nu_p \gg \Delta\nu_B$, which results in an increased Brillouin threshold, since the Brillouin power threshold is inversely proportional to the Brillouin-gain, according to

$$P_{\text{thres}} = \frac{21 A_{\text{eff}}}{g_B L_{\text{eff}}}$$ (8.19)

where A_{eff} is the effective core area and the effective interaction length is given by $L_{\text{eff}} = 1/\alpha[1 - \exp(-\alpha L)]$ with α the absorption coefficient in fibres and L the length of the fibre.

In order to reduce the SBS process, phase modulation of the optical field launched into the fibre has been proposed [8.34]. The technique is based on the mode-beating effect produced when the optical field comprises two discrete but closely spaced optical frequencies. This can be achieved using two single-frequency lasers operating at slightly different wavelengths or using a single laser which is arranged to operate on two longitudinal modes. In this way there is an increase of the spectral width of the pump beam and a corresponding decrease of the Brillouin-gain, according to the formula

$$g'_B = \frac{\Delta \nu_B^2}{2(\Delta \nu_B^2 + \Delta \nu_m^2)} g_B(\nu_B)$$ (8.20)

where g'_B is the peak Brillouin-gain, $g_B(\nu_B)$ is the Brillouin-gain coefficient for SBS which would be produced by a single-frequency laser, $\Delta \nu_m$ is the beat frequency of the two optical frequencies and $\Delta \nu_B$ is the spontaneous Brillouin scattering linewidth at the ambient temperature.

Experimental results in low-loss silica fibres, at 1.3 μm, have demonstrated more than a factor of ten suppression of the SBS gain. The explanation for the reduced Brillouin-gain is the following: if the optical phase reversals occur more frequently than the spontaneous acoustic dephasing, then the acoustic wave is unable to build up to a large amplitude and is not capable of giving rise to significant SBS.

In actual fibre communication systems the input light is modulated to carry information. Despite the very low threshold for SBS in low-loss silica fibres, coherent optical transmission systems can be designed to use modulation techniques which will eliminate any practical limitation on power levels and repeater spacing due to SBS.

Three commonly used formats are amplitude-shift keying (ASK), frequency-shift keying (FSK), and phase-shift keying (PSK) (figure 8.7). In the ASK modulation, the pump light is completely amplitude modulated. In the FSK modulation, the optical frequency is alternately changed, where one frequency denotes '1' and the other denotes '0'. In the PSK modulation, the phase of the pump light is alternately shifted from data '1' to '0'. The phase shift of the PSK modulation has been taken to be π, as used in most PSK systems.

In fibre-optic communication systems the spectral width $\Delta \nu_p$ increases considerably depending on the bit rate B at which the input signal is

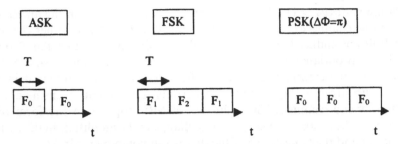

Figure 8.7. Schematic representation of fixed pattern ASK, FSK and PSK modulated lights (T is the pulse width and F is the frequency).

modulated. The amount by which the threshold power increases depends on the modulation format used for data transmission. The calculation of the Brillouin threshold under general conditions is quite complicated, as it requires a time-dependent analysis [8.35].

The SBS threshold for fixed pattern (1010...) ASK, FSK and PSK modulated lights are 2, 4 and 2.5 times higher, respectively, than the threshold for unmodulated light [8.36].

In the case of random pattern modulation, the pump and the Stokes waves cannot be expressed by a summation of discrete frequency components. The random pattern modulated pump light may be considered to be equivalent to a cw light that has a spectrum spread corresponding to one caused by random pattern modulation. The Brillouin threshold for the random pattern PSK format increases almost linearly with the bit rate B as:

$$P_{th}^{PSK} = \frac{B + \Delta\nu_B}{\Delta\nu_B} P_{th}^{cw} \tag{8.21}$$

which is quite different from the cases of ASK and FSK modulated lights [8.36]. For bit rates $B > 1\,Gb/s$ the SBS threshold is increased by more than one order of magnitude. At such high bit rates the threshold power even for ASK and FSK formats may increase substantially [8.31].

By using rare-earth-doped fibre amplifiers, transmitted powers can be boosted to a few hundred milliwatts. Such systems use semiconductor lasers with a narrow linewidth in combination with external modulation and SBS presence, which may cause major problems. The problem is solved by increasing the laser linewidth through sinusoidal modulation. For a 200 MHz modulation frequency the gain is reduced by a factor of 10, assuming $\Delta\nu_B = 20\,MHz$. Input power levels of up to 100 mW or more can be used by employing such a technique without suffering from SBS-induced degradation of the system performance [8.37].

In the case of multi-channel lightwave systems, SBS can lead to crosstalk if the fibre link supports channels in both directions and if the frequency difference between two counter-propagating channels happens to match the

Brillouin shift ($\nu_B = 11\,\text{GHz}$ at $1.55\,\mu\text{m}$). Crosstalk results in the amplification of one channel at the expense of the power carried by the other channel. The Brillouin-induced crosstalk is easily avoided by selecting the channel spacing appropriately because of the narrow frequency range ($\sim 100\,\text{MHz}$) over which the crosstalk can occur. The situation is quite different in the case of subcarrier multiplexing which permits simultaneous transmission of multiple video channels. The influence of SBS appears in the form of a decrease in the average optical power, a change in the amplitude in the carrier frequency and the appearance of the distortion components [8.38].

An additional limitation of SBS that may affect a multi-channel communication system is due to the SBS gain dependence on the frequency separation between two pump modes [8.39]. Such dependence may have influence on frequency-division-multiplexing optical communication systems, in which many frequency channels propagate simultaneously in an optical fibre. If the frequency separation between the channels is smaller than the Brillouin linewidth, the Brillouin-gain increases, resulting in a decrease in the maximum transmitted power. Moreover, under certain conditions, the relative phase between the input and scattered waves can double the SBS gain. Thus the frequency separation and the relative phase must be taken into account in designing optical communication systems.

8.4.2 Beneficial SBS applications to optical communication systems

Optical amplifiers are of great interest for optical communication systems. In long transmission systems optical amplifiers can replace expensive regenerative repeaters, and in multi-channel local area networks are used as signal boosters to compensate for split-off losses. The fibre scattering amplifiers use both Raman and Brillouin amplifiers. Fibre Raman amplifiers have a very broad bandwidth but require substantial amounts of pump power. Brillouin amplifiers, on the other hand, have a very narrow bandwidth and require small amounts of pump power.

All the applications of SBS in optical communication systems make use of narrow-bandwidth Brillouin amplification that can occur at relatively low pump powers. Typically, at $1.5\,\mu\text{m}$ wavelength the linewidth is only approximately $15\,\text{MHz}$, limiting the data rate to around $20\,\text{Mb/s}$.

The Brillouin-gain of an optical fibre can be used to amplify a weak signal whose frequency is shifted from the pump frequency by an amount equal to the Brillouin shift ν_B. However, the bandwidth of such an amplifier is generally below $100\,\text{MHz}$.

The unsaturated, single-pass amplifier gain G_A is given by

$$G_A = \exp(g_B P_0 L_{\text{eff}} / A_{\text{eff}}). \tag{8.22}$$

An exponential increase occurs only if the amplified signal power remains below the saturation power level.

Fibre Brillouin amplifiers are capable of providing 20–40 dB gain at a pump power of a few milliwatts. Their narrow bandwidth can be advantageous for some fibre-optic applications requiring selective amplification of only a narrow portion of the incident signal spectrum.

One such application is based on a method described in [8.40] in which the receiver sensitivity is improved by selective amplification of the carrier while leaving modulation sidebands unchanged. The amplification before detection of the carrier of a modulated optical signal by a narrowband quantum amplifier enhances the signal-to-noise ratio, thus enhancing the optical receiver sensitivity, particularly when the signal wave front is distorted. Under ideal conditions, the maximum sensitivity improvement is proportional to $G_A^{1/2}$ where G_A is the signal-pass amplifier gain given by equation (8.22). In a demonstration of this technique [8.41] the carrier was amplified by 30 dB more than the modulation sidebands even at a modulation frequency as low as 80 MHz. The limiting factor of the sensitivity improvement is the nonlinear phase shift induced by the pump if the difference between the pump and carrier frequencies does not exactly match the Brillouin shift.

The operation of a fibre Brillouin amplifier with electronically controlled bandwidth is reported in [8.42]. An increase of the fibre Brillouin amplifier bandwidth from 15 MHz to more than 150 MHz is demonstrated. Receiver sensitivity measurements at 10 and 90 Mb/s confirm the operation of the amplifiers and the scheme to extend the bandwidth. The receiver sensitivity was improved by 16 dB when a 10 Mb/s signal was transmitted over 30 km of fibre and amplified simultaneously by injecting 2.9 mW of pump power at the other fibre end. If the fibre-Brillouin amplifier is used as a preamplifier to the receiver, the amplifier performance would be limited by spontaneous Brillouin scattering.

Another application of the narrow line width of the Brillouin-gain profile consists of using it as a tuneable narrowband optical filter for channel selection in a densely packed multi-channel communication system [8.43]. The Brillouin amplifier can provide channel selectivity by amplifying the channel of interest and leaving nearby channels unamplified. This method does not impose severe restrictions on laser linewidths and bit rates using conventional direct-detection receivers. Using a tuneable colour-centre pump laser, whose frequency was tuned in the vicinity of the Brillouin shift, two 45 Mb/s channels were transmitted through a 10 km long fibre. Each channel could be amplified by 25 dB by using 14 mW pump power. Since $\Delta\nu_B < 100$ MHz typically, channels could be packed as close as $1.5\Delta\nu_B$ without introducing crosstalk from neighbouring channels.

Brillouin-gain has been used as a narrowband amplifier to simultaneously amplify and demodulate FSK signals, at bit rates up to 250 Mb/s [8.44]. Gain as much as a factor of 1000 was obtained for a pump power of 12 mW, using AlGaAs lasers for both pump and signal.

More information about phase conjugation by SBS in optical fibres can be found in [8.45–8.55].

8.5   Conclusions

In this chapter we have treated SBS in fibres both theoretically and experimentally and indicated the applications of the SBS process in optical fibre communication systems.

The analytical theoretical model presented for phase conjugation by SBS in optical fibres has described the exponential amplification of the Stokes intensity during propagation along the fibre, together with the decomposition on the system of fibre modes for multimode fibres.

High SBS reflectivities in glass multimode optical fibres (more than 80%) with core diameters from 25 µm up to 200 µm using pulse $\sim$30 ns, at 1.06 µm wavelength. Low SBS threshold $\sim$300 W can be obtained for the 25 µm fibre. High power densities up to $1\,\text{GW/cm}^2$ do not destroy these fibres. Pulse shaping and compression effect occur in fibre.

Finally, harmful and beneficial SBS applications to optical communications have been discussed.

More information about phase conjugation by SBS in optical fibres can be found in [8.45–8.55].

References

8.1 Ippen E P and Stolen R H 1972 *Appl. Phys. Lett.* **21** 539
8.2 Agrawal G P 1995 *Nonlinear Fiber Optics* 2nd edition (San Diego: Academic Press) p 370
8.3 Zeldovich B Y 1998 *Overview of optical phase conjugation* Technical Digest, Summaries Conference on Lasers and Electro-Optics, Conference Edition **6** 559 Opt. Soc. Am., Washington
8.4 Eichler H J, Haase A, Liu B and Mehl O 1998 *Laser-Physics* **8** 769
8.5 Kuzin E A, Petrov M P and Fotiadi A A 1994 Phase conjugation by SMBS in optical fibers in *Optical Phase Conjugation* ed M Gower and D Proch (Berlin: Springer) p 74
8.6 Eichler H J, Kunde J and Liu B 1997 *Opt. Commun.* **139** 327
8.7 Eichler H J, Kunde J and Liu B 1997 *Opt. Lett.* **23** 834
8.8 Eichler H J, Liu B, Haase A, Mehl O and Dehn A 1998 *Proc. SPIE* **3263** 20
8.9 Eichler H J, Haase A and Mehl O 1998 *Proc. SPIE* **3264** 9
8.10 Jun Chen, Zhi Hong, Chenfang Bao, Wenfa Qiu and Xiuping Wang 1998 *Proc. SPIE* **3549** 86
8.11 Dehn A, Eichler H J, Haase A, Mehl O and Schwartz J 1998 *Proc. SPIE* **3403** 65
8.12 Eichler H J, Haase A, Kunde J, Liu B and Mehl O 1997 *Power scaling of solid state lasers over 100 W with fiber phase conjugators* Solid State Lasers: Materials and Applications, Sino-American Topical Meeting 168, Technical Digest, Optical Society of America, Washington

8.13 Heuer A and Menzel R 1998 *Opt. Lett.* **23** 834

8.14 Heuer A and Menzel R 1999 *Stimulated Brillouin scattering in an internally tapered fiber* Proc. Int. Conf. LASERS'98 Soc. Opt. & Quantum Electron, McLean, Virginia p 1215

8.15 Harrison R G, Kovalev V I, Weiping Lu and Dejin Yu 1999 *Opt. Commun.* **163** 208

8.18 Kovalev V I and Harrison R G 1999 *Opt. Commun.* **166** 89

8.17 Eichler H J, Dehn A, Haase A, Liu B, Mehl O and Rücknagel S 1999 *Proc. SPIE* **3267** 158

8.18 Eichler H J, Mehl O, Risse E and Mocofanescu A 2000 *Continuously pumped all solid-state laser system with fiber phase conjugation* Technical Digest CLEO 2000, San Francisco 7–12 May p 404

8.19 Rodgers B C, Russell T H and Roh W B 1999 *Opt. Lett.* **24** 1124

8.20 Moore G T 2001 *IEEE J. Quantum Electron.* **37** 81

8.21 Hellwarth R W 1978 *J. Opt. Soc. Am.* **68** 1050

8.22 Rae S, Bennion I and Cardwell M J 1996 *Opt. Commun.* **123** 611

8.23 Lehmberg R H 1983 *J. Opt. Soc. Am.* **73** 558

8.24 Hu P H, Goldstone J A and Ma S S 1989 *J. Opt. Soc. Am. B* **6** 1813

8.25 Chu R, Kanefsky M and Falk J 1994 *J. Opt. Soc. Am. B* **11** 331

8.26 Mocofanescu A, Mehl O and Eichler H J 2001 *Proc. SPIE* **4430** 476

8.27 Lee C C and Chi S 2000 *IEEE Photon Tech. Lett.* **12** 672

8.28 Mocofanescu A 2001 *Nonlinear Optics* **27** 471

8.29 Damzen M J and Hutchinson H 1983 *IEEE Quantum Electron.* **QE-19** 7

8.30 Smith R G 1972 *Appl. Opt.* **11** 2489

8.31 Agrawal G P and Dutta N K 1993 *Semiconductor Lasers* (New York: Van Nostrand Reinhold)

8.32 Bar-Joseph I, Friesem A A, Lichtman E and Waarts R G 1985 *J. Opt. Soc. Am. B* **2** 1606

8.33 Valley G C 1986 *IEEE Quantum Electron.* **QE-22** 704

8.34 Cotter D 1982 *Electron. Lett.* **18** 638

8.35 Cotter D 1982 *Electron. Lett.* **18** 504

8.36 Aoki Y, Tajima K and Mito I 1988 *J. Lightwave Technol* **LT-6** 710

8.37 Miyamoto Y, Kataoka T, Sano A, Hagimoto K, Aida A and Kobayashi Y 1994 *Electron. Lett.* **30** 797

8.38 Yoshinaga H 1993 *Electron. Lett.* **29** 1707

8.39 Lichtman E, Friesem A A, Waarts R G and Yaffe H H 1987 *J. Opt. Soc. Am. B* **4** 11397

8.40 Arnaud J A 1968 *IEEE Quantum Electron.* **QE-4** 893

8.41 Cotter D, Smith D W, Atkins C G and Wyatt R 1986 *Electron. Lett.* **22** 6711

8.42 Olsson N A and van der Ziel J P 1987 *J. Lightwave Technol.* **LT-5** 147

8.43 Chraplyvy A R and Tkach R W 1986 *Electron. Lett.* **22** 1084

8.44 Tkach R W, Chraplyvy A R, Derosier R M and Shang H T 1988 *Electron. Lett.* **24** 260

8.45 Watanabe S, Ishikawa B, Naito T and Chikama T 1994 *J. Lightwave Tech.* **12** 2139

8.46 Matthews S C and Rockwell D A 1994 *Opt. Lett.* **19** 1729

8.47 Andreyev N F, Khanzanov E A, Palashov O I and Pasmanik G A 1994 *J. Opt. Soc. Am. B* **11** 786

8.48 Heil M M 1993 *IEEE J. Quant. Electron.* **29** 562

8.49 Li H P and Ogusu K 1999 *Japan J. Appl. Phys.* **38** 6309

8.50 Hsu H and Li T N 2000 *Appl. Opt.* **39** 6528

8.51 Kovalev V I, Harrison R G and Scott A M 2000 *Opt. Commun.* **185** 185

8.52 Tondagoldstein S, Dolfi D, Huignard J P, Charlet G and Chazelas J 2000 *Electron. Lett.* **36** 944

8.53 Gogolla T and Krebber K 2000 *J. Lightwave Tech.* **18** 320

8.54 Ogusu K 2000 *J. Opt. Soc. Am. B* **17** 769

8.55 Kovalev V I and Harrison R G 2000 *Phys. Rev. Lett.* **85** 1879

Chapter 9

Laser resonators with SBS mirrors

9.1 SBS phase conjugate lasers

The main approach in SBS phase conjugate lasers is to use the SBS reflection as one of the cavity mirrors of a laser resonator, to form a so-called phase conjugate resonator. The motivation for incorporating the SBS mirror is for the production of high spatial quality radiation despite aberrations within the laser system. Due to the threshold nature of the SBS process, the laser normally requires a conventional 'start-up' cavity to initiate the SBS mirror. Some simple cavity configurations are shown in figure 9.1. The nonlinearity of the SBS process can also change the temporal characteristics of the laser, including producing self-Q-switching.

Solid-state lasers have been the area on which most of the phase conjugate activity, using SBS, has focused. There are physical reasons for the development of SBS phase conjugate solid-state lasers. Their operating characteristics are compatible with the use of SBS because they can generate typically $\sim$10–100 ns pulses and even with small energies per pulse (in the range 1–10 mJ) this is sufficient to reach the SBS threshold. The solid-state lasers operate in the near infrared spectral range ($\sim$1 μm wavelength) where many materials offer high Brillouin-gain, reasonably short acoustic response times and low absorption. The broad gain bandwidth of gain media such as Nd:YAG (90 GHz) can easily accommodate the SBS shift which is typically 1–5 GHz in this spectral range. A comprehensive review of solid-state lasers using phase conjugation is found in [9.1].

Phase conjugate lasers have been demonstrated using many of the most common solid-state materials, including ruby, Nd:glass, Nd:YAG, Nd:YLF and Cr:Nd:GSGG. Many media have been utilized, including gases (CH_4, N_2, and SF_6), a variety of liquids (CS_2, CCl_4 and $TiCl_4$) and solid-state materials (quartz and recently optical fibres). Recent progress of SBS phase conjugate mirrors for a high power laser system is found in [9.2–9.5].

Phase conjugation by SBS offers a new alternative for energy scaling by allowing the coherent coupling of multiple parallel amplifiers, first reported

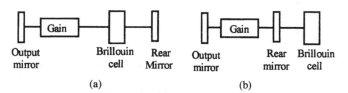

Figure 9.1. Experimental setup of a linear laser resonator with SBS mirror: (a) the Brillouin cell inside the resonator (b) the Brillouin cell outside the resonator.

by Basov [9.6]. This type of phase conjugate laser geometry has been utilized in scaling power of semiconductor and fibre lasers [9.7]. The two-cell Brillouin phase conjugate mirror is another approach for energy scaling because it not only minimizes the competing processes but offers a possibility of avoiding conjugation fidelity loss at high energies [9.8, 9.9].

Since diode pumping only reduces the heat load by up to a small factor relative to flash-lamp pumping for the same total stored energy, appreciable thermal effects still occur for producing high average powers. Hence phase conjugation will still be required to compensate the thermally-induced distortions in diode-pumped solid-state systems.

The first excimer laser phase conjugate MOPA with SBS mirror was reported by Gower and Caro [9.10] and implementation of phase conjugation in such lasers is reviewed by Gower [9.11]. Phase conjugation using SBS was extended to iodine lasers [9.12] and chemical lasers [9.13] and a phase conjugate chemical-laser MOPA using SBS was constructed first by Velikanov [9.14]. Generation of radiation of high brightness in gas lasers with an SBS mirror is found in [9.15] and phase conjugated lasers applied to X-ray generation is reported in [9.16].

Besides achieving a good beam quality, another feature provided by the SBS interaction is the self-Q-switching of the optical resonator within a laser oscillator [9.17]. The first Q-switching of the laser resonator using SBS was applied for ruby lasers [9.18] and then for Nd:YAG lasers. The approach usually adopted is to include the SBS medium inside a laser cavity with a secondary mirror employed to provide feedback at the beginning of the laser action, as shown in figure 9.1(a) [9.19, 9.20]. High intracavity intensities experienced by the SBS medium in this configuration can lead to poor spatial beam quality. This limits the repetition rate and peak power of the output. There has been reported also an alternative cavity configuration in which the SBS medium resides outside the conventional cavity [9.21–9.23], as illustrated in figure 9.1(b). The output beam is, in general, found to have a better beam quality and efficient energy extraction at higher repetition rate than the internal SBS case.

SBS was extended for its phase conjugating properties down to the UV range of wavelength, for excimer lasers [9.24] and into the IR domain, for Er lasers [9.25, 9.26].

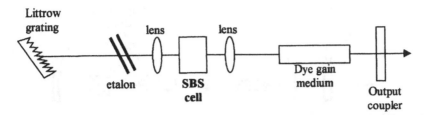

Figure 9.2. Schematic of laser with intracavity SBS cell.

9.2 Linear laser resonator with internal SBS cell

To illustrate the operation of an SBS resonator, we consider the system demonstrated in [9.27] of a dye laser with intracavity SBS cell as shown in figure 9.2. The dye gain medium was rhodamine 6G with a long pulse flash-lamp-pumping, length 45 cm, diameter 12 mm and a small-signal gain of 20. The conventional start-up cavity consisted of an output coupler with reflectivity of 30% and a back-reflector consisting of a Littrow grating with 1200 lines per millimetre. This grating reflector provides spectrally selective feedback with reflection arising from the first order of diffraction and could be used to tune the frequency of the broadband dye medium. An additional Fabry–Pérot etalon was incorporated to produce narrowband lasing and enhance the coherence length of the radiation to assist in the formation of the SBS process.

The intracavity SBS cell contained a high pressure gas $CClF_3$ and was placed in a focusing geometry at the centre of a pair of lenses of focal length 10 cm. The onset of cavity oscillation is between the output coupler and the rear grating. When there is sufficient intensity at the SBS cell, its reflectivity increases and becomes the dominant back-reflector. Indeed after a few SBS frequency shifts the cavity radiation moves out of the pass band of the Fabry–Pérot etalon and is isolated from the Littrow grating reflector. The output of this system consisted of a pulse of 800 ns duration and an energy of 450 mJ. The beam divergence was less than four times the diffraction limit. The corresponding output of the dye laser with no SBS cell and the back-mirror being a 100% reflector was 750 ns duration and 900 mJ output with a divergence $\sim$40 times the diffraction limit. The brightness of the SBS laser was $\sim$40 times larger than in the equivalent conventional cavity.

9.3 Linear laser resonator with external SBS mirror

In this section, an illustration of an experimental SBS laser system is made for the performance of a pulsed Nd:YAG laser system using external stimulated

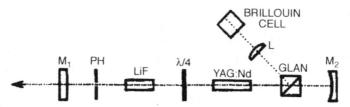

Figure 9.3. Experimental setup of linear laser resonator with external (side-arm) SBS mirror.

Brillouin scattering for Q-switching and phase conjugation of the cavity radiation. An example of a simple experimental configuration for the linear laser resonator with external SBS mirror is shown in figure 9.3 [9.28]. A concave back-mirror M_2 (3 m radius of curvature) with high reflectivity and a plane output coupling mirror M_1 with $R = 20\%$ defines a stable linear resonator whose length is 1.1 m. The gain element is a Nd:YAG rod with 6 mm diameter and 80 mm length, pumped by a flashlamp in a diffuse, liquid-cooled chamber and providing laser action at wavelength 1.06 µm. A quarter-wave plate–Glan polarizer combination provides an easy way to optimize the output energy coupling to a side-arm SBS cell. The quarter-wave plate is set to allow sufficient transmission for the linear cavity to reach threshold. Once lasing begins the coupling into the side-arm initiates SBS and establishes a phase conjugate resonator. The SBS reflectivity can rise to a high value and dominate the laser, compared with the conventional cavity which serves as a start-up cavity.

The Brillouin cell ($L = 10$ cm) is filled with carbon disulphide, CS_2, and is placed behind a lens (focal length $= 5$ cm) outside the linear cavity. CS_2 is chosen due to its large gain for SBS (0.06 cm/MW). This cell acted both as a phase conjugate mirror and as a Q-switch due to the transient increase in its reflectivity after the initial build-up of radiation in the start-up cavity. The pinhole (PH) selected the transverse modes by limiting the Fresnel number of the cavity.

9.4   Ring laser resonator with SBS mirror

For the linear configurations, each round trip the laser beam travels within the cavity will generate a downshifted Stokes beam and therefore will increase the bandwidth of the laser emission. A ring configuration allows the potential of single frequency operation from the resonator. In the ring configuration, the Q-switching regime is obtained as a start-up ring oscillator with two-pass amplifier system [9.29, 9.30]. The ring resonator geometry has been used for a dye laser with SBS Q-switching mirror [9.31, 9.32] and for solid-state (YSGG:Cr^{3+} crystal) in [9.33]. Besides carbon disulphide, which

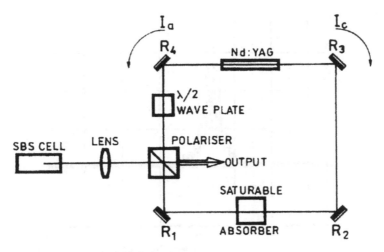

Figure 9.4. The experimental SBS ring laser resonator.

is a common nonlinear medium used for achieving SBS, other nonlinear media have been used including fluorocarbon [9.34] and L-arginine phosphate monohydrate crystal [9.35]. Processes limiting the output energy of Nd^{3+} YAG laser with SBS mirror are found in [9.36].

An experimental ring setup is depicted schematically in figure 9.4. Four plane mirrors with reflectivity $R \approx 100\%$ and the Nd:YAG rod with a small signal single pass gain of $\sim$100 defined the ring cavity, whose length was 1 m. The Nd:YAG rod with a 6 mm diameter and a 90 mm length was pumped by a flashlamp in a diffuse, liquid-cooled chamber. The ring cavity radiation at 1.06 µm was out-coupled by a half-wave plate–polarizer combination. This radiation was focused by a lens with a focal length of 10 cm into a Brillouin cell ($L = 10$ cm) filled with CS_2, placed outside the ring cavity. The Brillouin backscattering radiation returned to the ring resonator.

The energy entering the Brillouin cell was about 120–130 mJ in the form of relaxation oscillation pulses with a duration of the order of 200–300 ns. A saturable absorber (Kodak liquid saturable absorber for Nd:YAG laser radiation) could be used as a weak Q-switch to increase pulse energy and to enhance the initiation of the SBS process.

Output (I_0 from the clockwise radiation and I_c from the ring laser) were s-polarized and incident on the cell containing CS_2. Reflection from the SBS cell re-injected the output back into the ring where it contributed to the flux of the anticlockwise radiation, I_a. Because the SBS interaction preserves the polarization state, the re-injected radiation was also s-polarized but had a frequency which corresponded to the first Brillouin Stokes shift $\omega_1 = \omega_0 - \omega_B$, where ω_0 is the line-centre of the Nd:YAG crystal and ω_B is the acoustic frequency. After one complete round trip the s-polarized Stokes beam was strongly converted to p-polarization by the half-wave

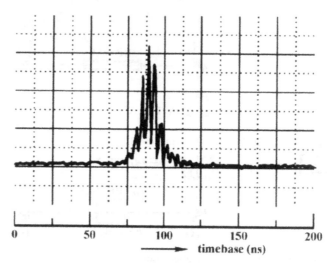

0   50   100   150   200

→ timebase (ns)

Figure 9.5. A typical output pulse of an SBS ring laser.

plate and then made a second round trip before reconversion back into s-polarization when it was rejected by the polarizer to produce the output (as shown in figure 9.4). The cavity therefore acts as an oscillator (clockwise direction)–two pass amplifier (anticlockwise direction) system.

The laser shown in figure 9.4 started in the free-running regime with a low Q factor determined by the losses of the cavity and of the saturable absorber. The SBS reflectivity of the cell increased rapidly with the incident wave intensity and the ring resonator with the Brillouin mirror reached a high Q factor. The output beam with energy of 40 mJ in a pulse of 15–20 ns duration was obtained at 2 Hz. A typical output pulse with a width (FWHM) of 15 ns is shown in figure 9.5. Under proper alignment of laser mirrors, and with an intracavity aperture of $d = 1.5$ mm the laser operated in the TEM$_{00}$ mode.

9.5 Theoretical modelling of passive Q-switching in SBS resonators

The passive Q-switching regime of the linear laser resonator with SBS mirror can be described by the rate equations [9.37]

$$\frac{\mathrm{d}q}{\mathrm{d}t} = q[Bn - \gamma_{\mathrm{nl}}(q)]$$

$$\frac{\mathrm{d}n}{\mathrm{d}t} = -qBn \tag{9.1}$$

$$n(0) = n_{\mathrm{th}}(1 + \zeta), \qquad q(0) = q_{\mathrm{sp}}$$

where q is the photon density, n is the population inversion, n_{th} is the threshold inversion, B is the Einstein coefficient and $\gamma_{nl}(q)$ represents the losses which are intensity dependent because of the SBS mirror reflectivity R. The initial population inversion $n(0)$ has been assumed above threshold by a factor $(1 + \zeta)$, $(\zeta \approx 2\%)$.

The threshold condition for laser oscillation is

$$IR_1 T e^{(\gamma_a - \alpha)l_r} T_G R_2 + R_G R_B I = I \qquad (9.2)$$

where I is the laser intensity in the resonator, R_1 is the reflectivity of the output mirror, $R_2 = 100\%$, T is the saturable absorber transmission, T_G and R_G are the transmission and the reflectivity of the Glan polarizer–waveplate combination, respectively, l_r is the Nd:YAG rod length, γ_a is the amplification coeficient and α represents the losses per unit length.

The nonlinear losses are

$$\gamma_{nl} = \left(\frac{l_r}{l_0}\right) c\alpha + \frac{c}{l_0} \ln \frac{1 - R_G R_B}{R_1 R_2 T T_G} \qquad (9.3)$$

where l_0 is the length of the resonator.

A simple steady-state theory of SBS gives the analytical solution for Brillouin reflectivity of the cell, R:

$$\frac{\exp(-G)}{R} = \frac{1 - R}{\exp[(1 - R)g_B L_{eff} I] - R} \qquad (9.4)$$

where $G \approx 25$ is related to the initial Stokes noise level, g_B is the Brillouin-gain and L_{eff} is an effective interaction length which depends on the radiation focusing inside the cell.

If we introduce the normalized variables

$$\xi = t/\tau$$
$$R(\xi) = I_s/I$$
$$y(\xi) = n(\xi)/n(0) \qquad (9.5)$$
$$x(\xi) = (I + I_s)g_B L_{eff}$$

where I_S is the Stokes intensity and $\tau = l_0/c$ is the round trip time of the resonator, the following system of equations was obtained:

$$\frac{dR(\xi)}{d\xi} = [A_0 y - G(R)]C(R)$$

$$\frac{dy(\xi)}{d\xi} = -B_0 y x \qquad (9.6)$$

$$\frac{dx(\xi)}{d\xi} = [A_0 y - G(R)]x.$$

The initial conditions are

$$R(0) \approx e^{-G} y(0) = 1, \qquad x(0) \approx e^{-G}. \tag{9.7}$$

The functions and constants used in system (9.6) are

$$A_0 = (l_0/c) Bn(0)$$

$$B_0 = \frac{2(l_0/c)B}{ch\nu g_B L_{\text{eff}}}$$

$$G(R) = \ln \frac{1 - R_G R_B}{R_1 R_2 T T_G} \tag{9.8}$$

$$C(R) = R(1 - R^2) \left[2R + \frac{1 - R^2}{\dfrac{1 + (1 - R)e^G}{1 + (1 - 2R)e^G} \ln[R + R(1 - R)e^G]} \right]^{-1}.$$

The results of the numerical simulations are presented in figure 9.6.

The theoretical model, which has considered the stationary approximation for the SBS interaction, proved to be satisfactory in explaining the Q-switch regime of the linear resonator. The predicted pulse duration is 25 ns for typical parameters as in the experiment for a linear resonator described previously.

The ring laser resonator can be treated in a similar way. More details about laser resonators with SBS mirrors can be found in the literature [9.38–9.47].

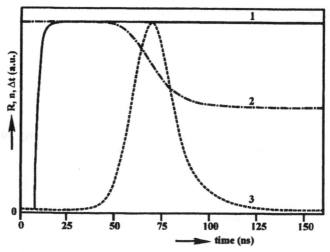

Figure 9.6. SBS reflectivity (1), population inversion (2), output pulse intensity (3) as a function of time.

9.6   Correction of aberration in laser amplifiers

A good beam quality of the laser light has been required for many scientific and technical applications of high-power solid-state lasers. Many applications require, or would benefit from, increased laser power. Unfortunately, intensive pumping of the laser amplifying medium leads to heating of the medium and degradation of laser performance. Heating results in temperature gradients and thermal expansion that produce refractive index changes so degrading beam quality, destabilizing the resonator and reducing laser efficiency.

Phase conjugation via SBS has been demonstrated to be a relatively simple and efficient method for dynamically correcting the aberrations in lasers and particularly solid-state lasers. Fundamental and applied aspects of phase conjugation using SBS are found in early reviews of the subject [9.48–9.50]. Phase conjugate lasers offer the possibility of scaling output energy with near diffraction limited beam quality.

Zel'dovich made the first demonstration of distortion compensation introduced through a passive aberrator using SBS [9.51]. Nosach [9.52] reported application of the same technique to a ruby master oscillator–power amplifier (MOPA) configuration, correcting for the distortions of the laser amplifier resulting in a large decrease of the output beam divergence. The MOPA configuration is shown schematically in figure 9.7. This configuration exploits the important fact that reciprocal phase distortions induced in an optical beam by a distorting medium can be compensated by reflecting the distorted beam with a phase conjugate mirror and passing it back through the distorting amplifier medium. A diffraction-limited beam quality produced by an oscillator is thereby restored following a phase conjugate reflection and a second pass through the amplifier. An optical isolator, such as a polarizer and quarter-wave plate combination, can be used to access the aberration-corrected beam.

The phase distortions produced by the intense pumping mechanism producing the inversion in the amplifying medium are described in chapter 4. In a solid-state laser rod it was shown that a thermally-induced lens is produced that has a power-dependent strength. In solid-state laser amplifiers, as well as phase distortion produced by thermally-induced refractive changes, a significant stress-induced birefringence is produced, especially in a rod geometry.

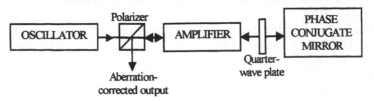

Figure 9.7. Experimental setup of a double-pass phase conjugate MOPA configuration.

The principal axes of the birefringence follow the cylindrical geometry of the rod and consist of radial and circumferential components. Radiation passing through the rod experiences a waveplate retardation by an amount dependent on the spatial position in the rod. The resultant amplified radiation is not a pure polarization state but is known as a depolarized beam. In this chapter, we shall mainly deal with the lensing aberrations which occur before the depolarization effects are important.

To reduce thermal loading, diode pumping can be applied, resulting in improved beam quality in comparison with flashlamp pumped laser systems [9.53]. In addition, advanced crystal geometry (slab [9.53–9.55] or disk laser [9.56]) can be applied to reduce the induced phase distortions. Recently, 1 kW output from a fibre-embedded disk laser was reported [9.57].

However, for average output powers above some tens of watts, the phase distortions lead to beam qualities far away from the diffraction limit. In these cases phase distortions have to be compensated with adaptive mirrors to facilitate near diffraction limited beam quality. Such mirrors can be realized by self-pumped phase conjugation based on SBS.

9.7   Pulsed MOPA systems with SBS mirrors

For high average power Nd-doped solid-state lasers, SBS is favourable for phase conjugating elements due to the absence of absorption inside the SBS medium.

In the case of amplifier systems the beam diameter can be adapted to the amplifier. Therefore a beam with good quality but low average power can be scaled up to the kW range.

To increase the average output power, parallel amplifier arrangements are preferable. In case of serial arrangements the output power is limited. (1) A single amplifier (the last one) suffers from the total output power and consequently large apertures are required to prevent damage to optical components. (2) the beam quality decreases after the first pass with an increasing number of amplification stages. Therefore the coupling efficiency drops (in the case of a fibre phase conjugator) or the SBS threshold strongly increases (when using a SBS cell with focusing geometry).

With regard to the advantages of a parallel amplifier setup, a solid-state laser system containing six amplifiers and a master oscillator was developed [9.58] (figure 9.8).

The laser is flashlamp pumped and pulsed at a typical repetition rate of 100 Hz. The master oscillator consists of a ring resonator and an additional etalon to increase the coherence length up to 23 cm. This guarantees a high reflectivity and fidelity from the phase conjugating mirrors. To reach the SBS threshold of about 20 kW a chromium-doped YAG-crystal is used as a passive Q-switch. Thus the oscillator emits per shot a train of 20 Q-switched

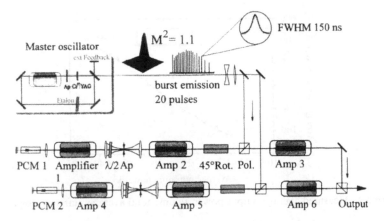

Figure 9.8. Experimental setup of a six-amplifier MOPA system with SBS phase conjugating mirrors based on SBS.

pulses, each with a width of 150 ns. The beam is diffraction limited at an average output power of 3.5 W.

The beam is divided into two parts, each of which is coupled into a serial arrangement of two amplifier rods. Each rod is excited in a diffuse pump chamber by two flashlamps with an average pumping power of up to 7 kW. To avoid depolarization of the beam during the amplification, the polarization preserving medium Nd:YALO has been used. Due to the anisotropic gain of YALO the optical isolation consists of a 45° Faraday rotator and a polarizer.

A phase conjugating mirror, which consists of an SBS cell filled with CS_2, compensates the phase distortions after the first amplifier pass. After the second amplification pass the beam profile of the oscillator is reproduced at the polarizer, used also as the output coupler. To increase the average output power the beams are amplified additionally in single-pass stages (amplifiers 3 and 6). This results in an acceptable decrease of beam quality. Using another polarizer, both beams are combined, resulting in one beam with a statistical polarization direction.

Due to the optical system between the rods and the SBS phase conjugating mirrors, the average output power can be varied over a wide range without changes of beam parameters.

Figure 9.9 shows the measured average power of the system against the total average pumping power. An output power up to 520 W was measured with a long term power stability (better than 3%). At an average output power of 400 W a five-times diffraction-limited beam quality is possible. The total stored power in the amplifier rods is 1200 W, therefore 43% of the stored power can be extracted with good beam quality [9.58].

In a similar multi-amplifier MOPA system (four amplifiers) but using a fibre phase conjugator instead of Brillouin cells, an average output power of

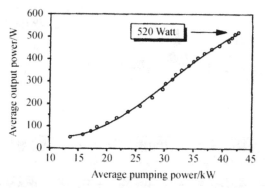

Figure 9.9. Measured average output power against total average pumping power.

315 W, at 2 kHz average repetition rate, with $M^2 = 2.6$ was reported [9.59]. The experimental setup is depicted in figure 9.10.

The beam is split in two at the thin film polarizer (TFP), each other perpendicular polarized parts, on to the two chains. By passing a Faraday rotator the polarization plane of each beam is rotated by an angle of 45° and is then amplified by the two serial amplifier arrangements. To compensate for astigmatic thermal lenses the amplifier rods (Nd:YALO) are rotated by 90° to each other and the polarization is matched again by a half-wave plate. After the single pass the beam is coupled into the fibre phase conjugator. The beam is phase conjugated by a multimode silica step index fibre with a core diameter of 200 μm, a numerical aperture of 0.22, and a length of 2 m. After an additional amplifier pass the initial beam quality is almost reproduced and the beam is rotated again by an angle of 45° after passing the Faraday rotator a second time. After the two passes through

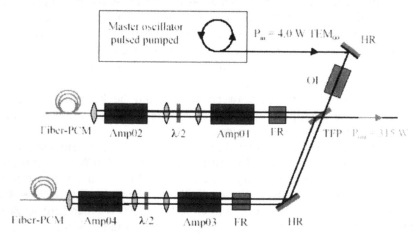

Figure 9.10. Experimental setup of a four-amplifier MOPA system with fibre phase conjugating mirrors based on SBS.

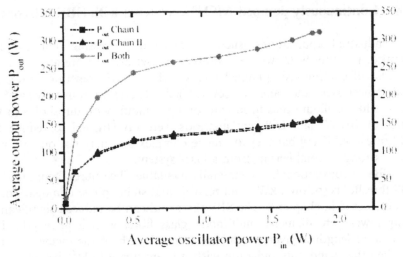

Figure 9.11. Measured average output power against average oscillator power.

the Faraday rotator the polarization plane of each beam is rotated by 90° and therefore could be extracted at the TFP. Here the amplified beams of both chains are superposed automatically due to the optical properties of the phase conjugated signal.

Figure 9.11 shows the measured output power as a function of the oscillator power.

The beam propagation factor M^2 was determined (according to the international standard ISO/CD 11 146) to be smaller than 2.6 for both directions in space (figure 9.12) [9.59].

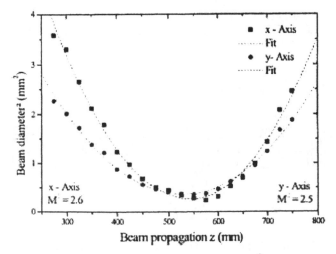

Figure 9.12. Determination of the beam propagation factor M^2.

9.8   Continuously pumped MOPA systems with SBS mirrors

For industrial applications, such as material processing, continuously pumped, repetitively Q-switched solid-state laser systems are attractive. In the case of continuously pumped lasers with $\sim 10\,$kHz repetition rate, the peak power is in the range of several kW. Therefore conventional SBS mirrors like Brillouin cells filled with organic liquids with thresholds in the range of 10 kW, fail. Different methods were investigated to reduce the SBS threshold [9.60] but they are more complex and may not operate well under strong thermal lensing from a laser system.

SBS in conventional, commercially available silica fibres facilitates low SBS thresholds (below 1 kW), and reliable and stable phase conjugation, as was discussed in chapter 8. The SBS threshold can be reduced significantly using lower core diameter multimode glass fibre, with long length. The interaction length is given by the coherence length of the incident beam. On the other hand only radiation with a beam quality (M^2) better than a limit, given as follows, can be coupled into the fibre: $M^2 \geq \pi D(NA/2\lambda)$, where λ is the wavelength and NA the numerical aperture of the fibre.

A continuously pumped four-pass amplifier arrangement with birefringence compensation, nearly diffraction-limited output with 30 W average power has been achieved at 10 kHz repetition rate [9.61, 9.62]. Figure 9.13 shows the experimental setup for such a system.

The master oscillator consists of a ring resonator and an arc lamp pumped Nd:YAG rod. Unidirectional operation is obtained with an external feedback mirror. An additional intracavity etalon reduces the bandwidth to

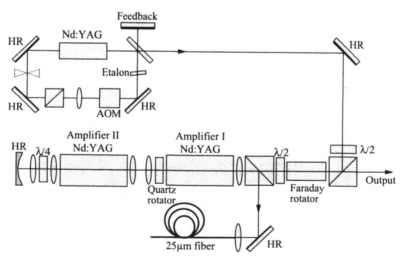

Figure 9.13. Continuously pumped MOPA system with four pass amplifier arrangement and fibre phase conjugator.

increase the possible interaction length inside the fibre. Q-switching with 10 kHz repetition rate is achieved using an acousto-optic modulator. The average output power is about 5.5 W. Pulse widths of 280 ns lead to pulse peak power of 2 kW.

The beam is phase conjugated in a multimode silica fibre with a core diameter of 25 μm, a numerical aperture of 0.22, and a length of 5 m.

The oscillator pulse energy of about 0.4 mJ is far below the saturation density of the used active medium Nd:YAG with a rod diameter of $\frac{1}{4}$ inch (6.35 mm). Therefore the extraction efficiency of the amplifier stages remains relatively low. However, the extraction efficiency can be improved by increasing the number of amplifier passes [9.63, 9.64]. After the first amplification pass, the beam is collimated and reflected using a conventional curved mirror. The polarization direction is rotated by 90°, passing the quarter-wave plate two times. During the second pass the beam is amplified again and then coupled into the fibre phase conjugator. After two additional amplifier passes the beam is extracted using optical isolation.

Figure 9.14 shows the measured output power as a function of the oscillator power. Applying the four-pass scheme the times diffraction-limit factor M^2 was determined to be smaller than 1.3 for both directions in space.

Power scaling was reported [9.59] using diode pumping of the YAG crystals and two further pumping chambers. After the second amplification pass the beam is reflected using a conventional HR-mirror and coupled in a

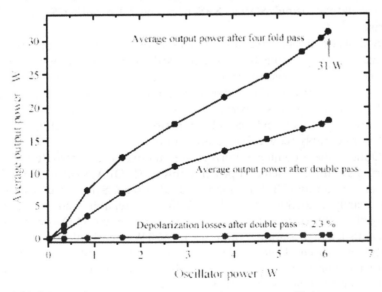

Figure 9.14. Measured average output power against average oscillator power.

further serial amplifier arrangement with birefringence compensation. After this amplification pass the beam is coupled into the fibre phase conjugator. The beam is phase conjugated in a multimode silica step index fibre with a core diameter of 100 μm, a numerical aperture of 0.22, and a length of 15 m. After three additional amplifier passes the beam is extracted with the help of optical isolation. Applying this scheme, an average output power up to 71 W is achieved at 10 kHz repetition rate with a nearly diffraction-limited beam quality. With further amplifier passes it is possible to increase the average output power to 120 W [9.59].

A cw Nd:YAG MOPA system, comprising a four-pass amplifier configuration with an SBS fibre phase conjugate mirror, is reported in [9.65]. The fibre phase conjugator (standard graded index silica multi-mode optical fibre of 50 μm core diameter and of 3.6 km length) allowed diffraction-limited output beam quality to be achieved. Maximal attainable gain of ∼1.9 for the system was obtained.

9.9   Conclusions

The SBS reflection can be used as one of the cavity mirrors of a laser resonator to form a so-called phase conjugate resonator. SBS mirrors can be used in oscillators and oscillator–amplifier arrangements.

Besides achieving a good beam quality, another feature provided by the SBS interaction is the self-Q-switching of the optical resonator within a laser oscillator. Experimental results and modelling of linear and ring resonators, using external SBS in a cell filled with CS_2, have been described.

Solid-state lasers have been the area on which most of the phase conjugate activity, using SBS, has focused. Solid-state laser sources are required for numerous applications in industry and science. Scaling of average output power while preserving a diffraction-limited beam quality results in high brightness operation. However, conventional laser systems suffer from thermally-induced phase distortions in the active medium, which considerably reduce their beam quality.

Optical phase conjugate mirrors are suitable to compensate for phase distortions in master oscillator power amplifier (MOPA) systems. Stimulated Brillouin scattering (SBS) in organic liquids (like CS_2) or in conventional, commercially available silica fibres facilitates reliable and stable phase conjugation. Pulsed pumped, passive Q-switched Nd:YALO systems which deliver an average output power up to 520 W, nearly diffraction-limited quality have been demonstrated. Furthermore a continuously pumped amplifier arrangement with nearly diffraction-limited output of 70 W average power has been achieved, at 10 kHz repetition rate.

References

9.1 Rockwell D A 1988 *IEEE J. Quantum Electron.* **24** 1124

9.2 Rockwell D A, Mangir M S and Ottusch J J 1993 *Int. J. Nonlinear Opt. Phys.* **3** 131

9.3 Eichler H J, Haase A and Mehl O 1999 High brightness solid state laser systems with phase conjugate mirrors in *Laser Resonators: Novel Design and Development* vol PM67 (SPIE Press) p 97

9.4 Ostermeyer M, Hodgson N and Menzel R 1998 *High-power fundamental mode Nd:YALO laser using a phase-conjugate resonator based on SBS* Summaries Conference on Lasers and Electro-Optics 1998 Opt. Soc. America, Washington, Technical Digest Series **6** 355

9.5 Yoshida H and Nakasuka M 1999 *Rev. Laser Engineering* **27** 95

9.6 Basov N G *et al* 1979 *Sov. J. Quantum Electron.* **9** 455

9.7 Rodgers B C, Russell T H and Roh W B 1999 *Opt. Lett.* **24** 1124

9.8 Vasil'ev A F and Yashin V E 1997 *Sov. J. Quantum Electron.* **17** 644

9.9 Crofts G J, Damzen M J and Lamb R A 1991 *J. Opt. Soc. Am. B* **8** 2282

9.10 Gower M C and Caro R G 1982 *Opt. Lett.* **7** 162

9.11 Gower M C 1984 *Proc. SPIE* **476** 72

9.12 Bessarab A V *et al* 1988 *Bull. Acad. Sci. USSR Phys. Ser.* **52** 333

9.13 Duignan M T, Feldman B J and Whitney W T 1987 *Opt. Lett.* **12** 111

9.14 Velikanov S D *et al* 1988 *Bull. Acad. Sci. USSR Phys. Ser.* **52** 553

9.15 Kulikov S M, Dudov A M, Dolgopolov Yu V, Kosyak E G, Kochemasov G G, Kirillov G A, Murugov V M, Ryadov A V, Smirnov A B, Sukharev S A, Shkapa A F and and Zykov L I 1992 *Proc. SPIE* **1628** 90

9.16 Hackel L A, Dane G B, Zapata L E and Hermann M R 1994 *Int. J. Nonlinear Opt. Phys.* **3** 137

9.17 Ostermeyer M, Mittler K and Menzel R 1999 *Phys. Rev. A* **59** 3975

9.18 Alcock A J and De Michelis C 1967 *Appl. Phys. Lett.* **11** 41

9.19 Meng H and Eichler H J 1991 *Opt. Lett.* 16 569

9.20 Eichler H J, Menzel R and Schumann D 1992 *Appl. Opt.* **31** 5038

9.21 Pashinin P P and Shklovsky E I 1991 *Laser Phys.* **1** 160

9.22 Agnesi A and Reali G C 1992 *Opt. Commun.* **89** 41

9.23 Lamb R A and Damzen M J 1993 *OSA Proc. Adv. Solid State Lasers* **15**

9.24 Eichler H J, Heinrich S and Schwartz J 1996 *Opt. Lett.* **21** 1909

9.25 Eichler H J, Liu B and Sperlich O 1997 *3μm Er:Cr:YSGG-laser with a SBS phase conjugating mirror* Solid State Lasers: Materials and Applications, Sino-American Topical Meeting, Technical Digest, Opt. Soc. America, Washington p 118

9.26 Eichler H J, Liu B and Sperlich O 1998 *Proc. SPIE* **3265** 75

9.27 Soan P J, Damzen M J, Aboites V and Hutchinson M H R 1994 *Opt. Lett.* **19** 783

9.28 Mocofanescu A, Babin V, Miclos S and Farcas A 1998 *Proc. SPIE* **3405** 45

9.29 Mocofanescu A, Udaiyan D, Damzen M J and Babin V 1997 *Ro. J. Phys.* **42** 177

9.30 Mocofanescu A, Udaiyan D, Damzen M J and Babin V 1995 *Proc. SPIE* **2461** 294

9.31 Case A D, Soan P J, Damzen M J and Hutchinson M H H 1992 *J. Opt. Soc. Am. B* **9** 374

9.32 Barrientos B, Aboites V and Damzen M J 1996 *Appl. Opt.* **35** 5386

9.33 Vokhnik O M and Terent'eva I V 1998 *Opt. Spectrosc.* **85** 797

9.34 Yoshida H, Kmetik V, Fujita H, Nakatsuka M, Yamanaka T and Yoshida K 1997 *Appl. Opt.* **36** 3739
9.35 Yoshida H, Nakatsuka M, Fujita H, Sasaki T and Yoshida K 1997 *Appl. Opt.* **36** 7783
9.36 Batishche S A, Kuzmuk A A and Malevich N A 1998 *Quantum Electron.* **28** 379
9.37 Siegman A E 1986 *Lasers* (Stanford: University Science Books)
9.38 Lamb R A 1996 *J. Opt. Soc. Am. B* **13** 1758
9.39 Kiryanov V, Aboites V and Ilichev N N 2000 *J. Opt. Soc. Am. B* **17** 11
9.40 Kim H S, Ko D K, Jung E C, Lim C, Lim G, Cha B H and Lee J 2000 *Opt. Lett.* **25** 399
9.41 Dane C B, Zapata L E, Neuman W A and Hackel L A 1995 *IEEE J. Quant. Electron.* **31** 148
9.42 Perrone M R and Yao Y B 1994 *IEEE J. Quant. Electron.* **30** 1327
9.43 Bose M, Aghamkar P and Sen P K 1992 *Phys. Rev. B* **46** 1395
9.44 Whitney W T, Duignan M T and Feldman B J 1992 *Appl. Opt.* **31** 699
9.45 Ayral J L, Montel J and Huignard J-P 1991 *Proc. SPIE* **1500** 81
9.46 Alimpiev S S, Bukreev V S, Vartapetov S K, Veselovskii I A, Kusakin V I, Likhanskii S V and Obidin A Z 1991 *Sov. J. Quant. Electron.* **21** 80
9.47 Whitney W T 1990 *J. Opt. Soc. Am. B* **7** 2160
9.48 Fisher R A ed 1983 *Optical Phase Conjugation* (New York: Academic)
9.49 Pepper D M 1985 Nonlinear optical phase conjugation in *Laser Handbook* vol 4 ed M L Stich and M Bass (Amsterdam: North-Holland)
9.59 Zel'dovich B Ya, Pilipetsky N F and Shkunov V V 1985 *Principles of phase conjugation* Springer Ser. Opt. Sci. 42 (Berlin: Springer)
9.51 Zel'dovich B Ya, Popovichev V I, Ragul'skii V V and Faizullov F S 1972 *JETP Lett.* **15** 109
9.52 Nosach O Yu, Popovichev V I, Ragul'skii V V and Faizullov F S 1972 *JETP Lett.* **16** 435
9.53 Schone W, Knoke S, Tunnermann A and Welling H 1997 *Efficient diode-pumped cw solid-state lasers with output powers in the kW range* CLEO'97 Technical Digest Series **11** talk CFE2
9.54 Kiriyama H *et al* 1998 *Proc. SPIE* **3264** 30
9.55 Dane C B, Zapata L E, Neuman W A, Norton M A and Hakel L A 1995 *IEEE J. Quantum Electron.* **31** 148
9.56 Giesen A, Hollemann G and Johannsen I 1999 Technical Digest CLEO '99 29
9.57 Ken-ichi Ueda 2002 CLEO '02 Long Beach Postdeadline paper talk CPDC4
9.58 Eichler H J, Haase A and Mehl O 1999 *SPIE Press* **PM67** 97
9.59 Riesbeck T, Risse E and Eichler H J 2001 *Appl. Phys. B* **73** 847; *High brightness all solid state laser systems with fiber phase conjugate mirrors* The XIV International symposium on gas flow and chemical lasers and high power lasers, Wroclaw (Poland) 2002 to be published in *Proc. SPIE*
9.60 Scott A M and Ridley K D 1989 *IEEE J. Quantum Electron.* **25** 438
9.61 Eichler H J, Dehn A, Haase A, Liu B, Mehl O and Rücknagel S 1998 *Proc. SPIE* **3265** 200
9.62 Eichler H J, Mehl O, Risse E and Mocofanescu A 2001 *Proc. SPIE* **4184** 179
9.63 Eichler H J, Mehl O and Eichler J 1999 *Proc. SPIE* **3613** 166
9.64 Andreev N F, Pasmanik G A, Pashinin P P, Sergeev S N, Sierov R V, Shklovskii E I and Yanovskii V P 1983 *Sov. J. Quantum Electron.* **13** 641

Chapter 10

Optical solitons in SBS

10.1 Optical solitons

The soliton was discovered by J S Russel, in hydrodynamics (1834). In 1964, Zabusky and Kruskal introduced the word soliton. Then, it was demonstrated that solitons could be present in all wave physics, including optics. In 1973, Hasegawa and Tappert showed theoretically that an optical pulse can form an envelope soliton in an optical fibre, and in 1980 Mollenauer demonstrated the effect experimentally [10.1, 10.2]. Optical solitons are studied intensely due to the distortion-less transmission in ultra-fast communication systems and to the many interesting nonlinear applications in information optics.

The optical soliton is a wave (envelope), which preserves its time–space shape by the dispersion–diffraction compensation by the optical nonlinearity of the propagation medium.

In a dispersive medium (with frequency dependent refractive index), the envelope of a modulated light wave (particularly, of light pulses) is distorted due to different velocities of its frequency components.

The time-dependent paraxial wave equation, in the presence of dispersion, can be written as [10.1–10.5]

$$i\left(\frac{\partial}{\partial z} + k'\frac{\partial}{\partial t}\right)E - \frac{k''}{2}\frac{\partial^2 E}{\partial t^2} = 0 \tag{10.1}$$

where the derivatives of the wavevector k, ($k' = \partial k/\partial \omega$, $k'' = \partial^2 k/\partial \omega^2$) with respect to frequency can be related to the group velocity, v_g, by the relations

$$v_g = \frac{\partial \omega}{\partial k} = \frac{1}{k'}, \qquad k'' = \frac{\partial}{\partial \omega}\left(\frac{1}{v_g}\right) = -\frac{1}{v_g^2}\frac{\partial v_g}{\partial \omega}. \tag{10.2}$$

Equation (10.2) provides the frequency dependence of the group velocity and consequently, k'' can be identified to the dispersion of the group velocity of the wave, often defined as $D = k''(2\pi c/\lambda^2)$. Owing to the fact that the envelope of the light wave propagates with the group velocity, it is convenient to represent the wave propagation in a coordinate system, which moves at the group velocity by

$$\xi = \varepsilon^2 z, \qquad \tau = \varepsilon(t - k'z) \tag{10.3}$$

where $\varepsilon = \Delta\omega/\omega$ is the relative width of the spectrum. The propagation equation becomes

$$i\frac{\partial E}{\partial \xi} - \frac{k''}{2}\frac{\partial^2 E}{\partial \tau^2} = 0. \tag{10.4}$$

In the Kerr nonlinear media (and SBS can be considered such), the refractive index depends on the light electric field as

$$n = n_0(\omega) + n_2|E|^2 \tag{10.5}$$

and the wave propagation equation can be written as

$$i\frac{\partial E}{\partial \xi} - \frac{k''}{2}\frac{\partial^2 E}{\partial \tau^2} + g\frac{|E|^2 E}{\varepsilon^2} = 0 \tag{10.6}$$

with $g = 2\pi n_2/\lambda$, which is named the nonlinear Schrödinger equation (considering the potential in the well-known equation proportional to $|E|^2$). Following Hasegawa [10.1], in order to derive a stationary solution of equation (10.6), one can use the normalized variables

$$q = \frac{\sqrt{g\lambda}}{\varepsilon}E, \qquad Z = \frac{\xi}{\lambda}, \qquad T = \frac{\tau}{(-\lambda k'')^{1/2}} \tag{10.7}$$

which lead to the following form of the propagation equation

$$i\frac{\partial q}{\partial Z} + \frac{1}{2}\frac{\partial^2 q}{\partial T^2} + |q|^2 q = 0. \tag{10.8}$$

A localized, stationary (in Z), single-humped solution of equation (10.8) can be written with real and imaginary parts as [10.1]

$$q(T, Z) = A \operatorname{sech} A(T + V_s Z - \theta_0) \exp\left\{-iV_s T + \frac{i}{2}(A^2 - V_s^2)Z - i\theta_1\right\} \tag{10.9a}$$

where A is the amplitude and the inverse width of the soliton, θ_0 and θ_1 are the phase constants and V_s is the velocity of the soliton pulse propagation (actually a deviation from the group velocity, which is independent of the soliton amplitude, A). We can remark that the soliton amplitude is inversely proportional to the soliton width. This solution exists in the region of anomalous dispersion ($k'' < 0$) only. In the region of normal dispersion ($k'' < 0$), the solution is called dark soliton [10.1–10.5].

The simplest soliton solution, which can be derived from equation (10.9a) with proper normalization, phase constants and with negligible relative soliton velocity, is

$$q(T, Z) = \text{sech}(T) \exp\left(\frac{i}{2}Z\right). \tag{10.9b}$$

In this equation, it is possible to identify $T = \tau/t_L$ and $Z = \beta''/t_L^2$, where τ is the time in the time frame moving with the group velocity of the pulse and where t_L is the pulse width. The soliton period (related to its phase and corresponding to the distance at which the pulse doubles its width in propagation through linear media) is $Z = \pi/2$. The solitons are waves, which propagate in nonlinear media without changing their form due to a compensation of the dispersion effects by the nonlinear effects. Thus, a certain power is required in order to produce this compensation:

$$P = t_L^{-2}(\beta'' A_{\text{eff}}/n_2 k_0) \tag{10.10}$$

where A_{eff} is the effective area of light beam and k_0 is the free space wavevector. If we consider, as often invoked, the soliton formation in optical fibres with a typical dispersion of $15 \, \text{ps} \, \text{nm}^{-1} \, \text{km}^{-1}$ and a nonlinear refractive index $n_2 = 3.2 \times 10^{-16} \, \text{cm}^2/\text{W}$, at $\lambda = 1.5 \, \mu\text{m}$, the principal parameters required for maintaining the pulse width of t_L [ps] are [10.4] $P[\text{W}] = 6.9/t_L^2$, $E[\text{pJ}] = 13.6/t_L$ and $z[\text{km}] = 0.009 t_L^2$.

10.2 Optical solitons in SBS

A large class of nonlinear processes was investigated using the inverse problem in the scattering theory [10.6–10.8]. The study of the optical non-linear processes using the inverse problem in the scattering theory revealed the physical conditions for the existence of optical solitons and allowed the description of their dynamics (the evolution in nonlinear media, collisions, interaction between several solitons etc). Calogero and Degasperis [10.6] described the spectral analysis of the nonlinear equations of evolution that result from the inverse problem in the scattering theory. Ablowitz and Segur [10.7] gave a rigorous characterization of solitons in the inverse method in the scattering theory. Novikov *et al* [10.8] analysed the existence of solitons in the interaction process of several waves in nonlinear media.

SBS could be characterized by a third-order nonlinear susceptibility. Formally, a nonlinear Schrödinger equation as in equation (10.6) can be written and an SBS soliton solution could be considered. However, the non-linear susceptibility was previously calculated in some particular cases only and cannot describe the full complexity of this problem. Agrawal [10.3] suggested that the coupled-amplitude SBS (three) equations, written for the time scale shorter than the phonon lifetime (in the order of $10 \, \text{ns}$), could admit coupled soliton solutions. These solutions were similar to those obtained in stimulated Raman scattering (SRS) with cross-phase modulation. Agrawal related his assertion to the experimental work of Montes *et al* [10.25]. These authors studied the dynamics of mode-locked ring lasers with a Brillouin mirror and found out that, for a small SBS reflectivity, a stable train of (longitudinal) soliton impulses could occur in the cavity. The temporal profile of the Stokes solitons and their stability were studied experimentally and by numerical simulations (with the pump and Stokes wave equations only), using three control parameters: the SBS reflectivity, the normalized Brillouin-gain, $\sigma_B = g_B I_0 L$ and the number of the longitudinal modes in the SBS material (an optical fibre).

The aim of this chapter is a rigorous treatment of the complete SBS set of equations, including the acoustic one (in a more general case with respect to the models used in previous chapters), in order to demonstrate and to characterize the temporal Stokes solitons. We process the SBS equations using the derivatives along the characteristic directions of the wave equation solution [10.9] and we look for analytical soliton solutions in the optical and in the acoustic fields.

We shall first analyse the soliton, which may arise from the dispersion compensation by the SBS nonlinearity, its duration and velocity in the nonlinear medium. This soliton is defined here as *compensation soliton*.

Then, we analyse the SBS equations in the case of very low dispersion and absorption ($k(\omega) \approx 0; \alpha \approx 0$). In this case, another type of SBS soliton may result from a condition imposed on the SBS nonlinear equation system, written in the phase space. This SBS soliton is called by us *topological soliton*. Sagdeev *et al* [10.10] have done a tentative study to identify similar solitons in hydrodynamics.

In the SBS process, as presented in chapter 1 (equations (1.18), (1.19) and (1.21)), the incident light, the scattered light and the acoustic plane wave evolutions on the propagation axis (z), are represented by

$$\frac{\partial E_L}{\partial z} + \frac{n}{c}\frac{\partial E_L}{\partial t} + \frac{1}{2}\alpha E_L = \frac{i\omega_L}{4cn}\frac{\gamma_e}{\rho_0} E_S \rho$$

$$-\frac{\partial E_S}{\partial z} + \frac{n}{c}\frac{\partial E_S}{\partial t} + \frac{1}{2}\alpha E_S = \frac{i\omega_S}{4cn}\frac{\gamma_e}{\rho_0} E_L \rho^* \qquad (10.11)$$

$$\frac{\partial \rho}{\partial t} + \frac{\Gamma_B}{2}\rho = \frac{i\gamma_e\varepsilon_0 K}{4v} E_L E_S^*$$

where E_L is the pump (incident) electric field, E_S is the scattered electric field, ρ is the density variation due to the interaction with the laser (optical) field, ρ_0 is the density of the propagation medium, v and K are the hypersound velocity and wavevector, respectively, Γ_B is the Brillouin linewidth, γ_e is the electrostrictive coefficient dielectric and ε_0 is the permittivity of free space.

The electric and the magnetic fields are linearly polarized and the conservation relations hold (in back-scattering):

$$K = K_L + K_S \cong 2K_L, \qquad \omega = \omega_L - \omega_S. \qquad (10.12)$$

In order to simplify the SBS equations, one can introduce some new variables, with the physical meaning of phases of the interacting waves:

$$\varphi_L = \omega_L t + K_L z = \xi_L K_L$$
$$\varphi_S = \omega_S t - K_S z = \xi_S K_S \qquad (10.13)$$
$$\varphi = \omega t - K z = \xi_f K$$

which define the three characteristic directions $\{\xi_L, \xi_S, \xi_f\}$ [10.9, 10.11, 10.19].

Using the derivatives along the characteristic directions of integral solutions of the corresponding wave equations ([10.9] and appendix 1), the SBS equation system (10.11) may take a form similar to that from equation (5.46), without transverse effects:

$$\frac{\partial E_L'}{\partial \varphi_L} = -\frac{\alpha'}{2} E_L' - \frac{\partial}{\partial \varphi_L}(E_S' E_{ac}')$$

$$\frac{\partial E_S'}{\partial \varphi_S} = -\frac{\alpha'}{2} E_S' + \frac{\partial}{\partial \varphi_S}(E_L' E_{ac}') \qquad (10.14)$$

$$\frac{\partial E_{ac}'}{\partial \varphi} = -2A E_{ac}' + \sigma_B(E_L' E_S')$$

where the normalized amplitudes of the pump, Stokes and acoustic fields are (in Gaussian units)

$$E_L' = \sqrt{\frac{cn}{8\pi I_0}} E_L, \qquad E_S' = \sqrt{\frac{cn}{8\pi I_0}} E_S, \qquad E_{ac}' = \frac{\pi \gamma_e}{n^2}\left(\frac{\Delta \rho}{\rho_0}\right) \qquad (10.15)$$

and I_0 is the maximum pump intensity, $\alpha' = \alpha/2K_L = \alpha/2K_S \approx \alpha/K$ is the normalized absorption coefficient, $A = 2\omega/\Gamma_B$ is the gain of the acoustic field, $\sigma_B = g_B L_B I_0$ is the normalized Brillouin-gain and L_B is the interaction length.

In order to solve analytically equations (10.14), one can define the initial conditions

$$E_L'(\varphi_L)|_{\varphi_L = \varphi_{L_0}} = E_{L0}'(\varphi_{L_0})$$

$$E_S'(\varphi_S)|_{\varphi_S = \varphi_{S_0}} = E_{S0}'(\varphi_{S_0}) \qquad (10.16)$$

$$E_{ac}'(\varphi)|_{\varphi = \varphi_0} = E_{ac0}'(\varphi)$$

where

$$\varphi_L|_{z=0} = \varphi_{L_0}, \qquad \varphi_S|_{z=0} = \varphi_{S_0}, \qquad \varphi|_{z=0} = \varphi_0. \qquad (10.17)$$

With the scalar transformation

$$\hat{E}_{\mathrm{L}} = \sqrt{L_{\mathrm{B}}/L'_{\mathrm{B}}}\,E'_{\mathrm{L}}, \qquad \hat{E}_{\mathrm{S}} = \sqrt{L_{\mathrm{B}}/L'_{\mathrm{B}}}\,E'_{\mathrm{S}} \qquad (10.18)$$

which uses L'_{B} as a normalization length (in order to bring the light fields to the scale of the acoustic wavelength), the system (10.14) becomes

$$\frac{\partial \hat{E}_{\mathrm{L}}}{\partial \varphi_{\mathrm{L}}} = -\frac{\alpha'}{2}\hat{E}_{\mathrm{L}} - \frac{\partial}{\partial \varphi_{\mathrm{L}}}(\hat{E}_{\mathrm{S}}E'_{\mathrm{ac}})$$

$$\frac{\partial \hat{E}_{\mathrm{S}}}{\partial \varphi_{\mathrm{S}}} = -\frac{\alpha'}{2}\hat{E}_{\mathrm{S}} + \frac{\partial}{\partial \varphi_{\mathrm{S}}}(\hat{E}_{\mathrm{L}}E'_{\mathrm{ac}}) \qquad (10.19)$$

$$\frac{\partial E'_{\mathrm{ac}}}{\partial \varphi} = -(2A)E'_{\mathrm{ac}} + \sigma_{1\mathrm{B}}\hat{E}_{\mathrm{L}}\hat{E}_{\mathrm{S}}, \qquad \sigma_{1\mathrm{B}} = g_{\mathrm{B}}L'_{\mathrm{B}}I_0.$$

We are looking for a solution of the system (10.19) with the following form

$$\hat{E}_{\mathrm{L}} = x_1\,\mathrm{e}^{\mathrm{i}\varphi_{\mathrm{L}}} + x_1^*\,\mathrm{e}^{-\mathrm{i}\varphi_{\mathrm{L}}}$$

$$\hat{E}_{\mathrm{S}} = y_1\,\mathrm{e}^{\mathrm{i}\varphi_{\mathrm{S}}} + y_1^*\,\mathrm{e}^{-\mathrm{i}\varphi_{\mathrm{S}}} \qquad (10.20)$$

$$E'_{\mathrm{ac}} = z_1\,\mathrm{e}^{\mathrm{i}\varphi} + z_1^*\,\mathrm{e}^{-\mathrm{i}\varphi}.$$

Introducing equations (10.20) in equations (10.19), one can obtain

$$\frac{\partial}{\partial \varphi_{\mathrm{L}}}(x_1 + y_1 z_1) + \mathrm{i}(x_1 + y_1 z_1) = -\frac{\alpha'}{2}x_1,$$

$$\frac{\partial}{\partial \varphi_{\mathrm{S}}}(y_1 - x_1 z_1^*) + \mathrm{i}(y_1 - x_1 z_1^*) = -\frac{\alpha'}{2}y_1, \qquad (10.21)$$

$$\frac{\partial z_1}{\partial \varphi} + \mathrm{i}z_1 = -(2A)z_1 + \sigma_{1\mathrm{B}}x_1 y_1^*,$$

where x_1^* is the complex conjugated of x_1.

The system (10.21) can be put in a parametric form by 'projecting' the evolution along $[\varphi_{\mathrm{L}}]$ and $[\varphi_{\mathrm{S}}]$ characteristics on the acoustic field characteristic $[\varphi]$. The configuration of the characteristics φ_{L}, φ_{S}, φ, in the $\{z, t\}$ plane, is presented in figure 10.1.

From figure 10.1, one can find (observing that $c/n \gg v/\sqrt{\gamma}$, [10.24])

$$\frac{\mathrm{d}\varphi_{\mathrm{L}}}{\mathrm{d}\varphi} = \frac{\omega_{\mathrm{L}}}{\omega}\delta, \qquad \frac{\mathrm{d}\varphi_{\mathrm{S}}}{\mathrm{d}\varphi} = \frac{\omega_{\mathrm{L}}}{\omega}\delta \qquad (10.22)$$

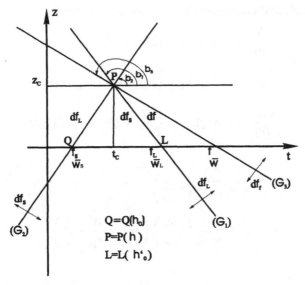

Figure 10.1. The characteristics in the space-time plane: Γ_1: $\varphi_L = \omega_L t + K_L z$; Γ_2: $\varphi_S = \omega_S t - K_S z$; Γ_3: $\varphi = \omega t + Kz$.

with

$$\delta = \left| \frac{1 + \left(\dfrac{n}{c}\right)\left(\dfrac{dz}{dt}\right)_{\varphi_L}}{1 + \dfrac{\sqrt{\gamma}}{\upsilon}\left(\dfrac{dz}{dt}\right)_{\varphi}} \right| = \left| \frac{1 + \left(\dfrac{n}{c}\right)\left(\dfrac{dz_c - z}{dt}\right)_{\varphi_S}}{1 + \dfrac{\sqrt{\gamma}}{\upsilon}\left(\dfrac{dz}{dt}\right)_{\varphi}} \right| . \qquad (10.23)$$

Using equations (10.22), equations (10.21) can be written as

$$\frac{\partial}{\partial \varphi}(x_1 + y_1 z_1) + i\left(\delta\frac{\omega_L}{\omega}\right)(x_1 + y_1 z_1) = -\frac{\alpha'}{2}\left(\delta\frac{\omega_L}{\omega}\right)x_1$$

$$\frac{\partial}{\partial \varphi}(y_1 - x_1 z_1^*) + i\left(\delta\frac{\omega_L}{\omega}\right)(y_1 - x_1 z_1^*) = -\frac{\alpha'}{2}\left(\delta\frac{\omega_L}{\omega}\right)y_1 \qquad (10.24)$$

$$\frac{\partial}{\partial \varphi}z_1 + i z_1 = -(2A)z_1 + \sigma_{1B}x_1 y_1^* .$$

Furthermore, a nonlinear transform can be used to bring the system (10.24) into the real form [10.12, 10.15, 10.24]. In this case, the solutions of the real (differential hyperbolic, quasi-linear) system are implicit functions of Riemann invariants, $\{N_1(\eta), N_2(\eta), N_3(\eta)\}$, which are associated to the system (10.24) and consequently, are related as follows:

$$\frac{\partial N_1}{\partial \eta} = -(4\gamma_1)N_1 + (2\gamma_1)N_2 N_3 - \gamma_2$$

$$\frac{\partial N_2}{\partial \eta} = -(4\gamma_1)N_2 - (2\gamma_2)N_3 - (8\gamma_1)N_1 N_3 \qquad (10.25)$$

$$\frac{\partial N_3}{\partial \eta} = +(4\gamma_1)N_3 + \frac{1}{2}N_2 + 2N_1 N_3 - 2N_2 N_3^2$$

where

$$\gamma_1 = \frac{\alpha'}{8\sigma_{1B}}\left(\delta\frac{\omega_L}{\omega}\right)$$

$$\gamma_2 = \frac{\alpha'}{8\sigma_{1B}}\left(\delta\frac{\omega_L}{\omega}\right)\left(\frac{A}{\sigma_{1B}} + \frac{\alpha'}{\sigma_{1B}}\delta\frac{\omega_L}{\omega}\right) \qquad (10.26)$$

and

$$\eta = 2\sigma_{1B}\varphi|_{\varphi_S} = 2\sigma_{1B}\frac{\omega}{\omega_S\delta}[\omega_S t - k_S(z_c - z)]. \qquad (10.27)$$

The invariants $\{N_1(\eta), N_2(\eta), N_3(\eta)\}$ are functions of the normalized intensities of optical pump, Stokes and acoustic waves, $x_1 x_1^*$, $y_1 y_1^*$ and $z_1 z_1^*(\eta)$, respectively. The following equations can be written [10.24]

$$x_1 x_1^* - y_1 y_1^* = \frac{4(1 - N_3^2)}{1 + 3N_3^2}\left(N_1 + \frac{\gamma_2}{4\gamma_1} - \frac{N_2 N_3}{1 - N_3^2}\right)$$

$$(x_1 x_1^*)(y_1 y_1^*) = \left[\frac{N_2}{1 - N_3^2} + \frac{4N_3}{1 + 3N_3^2}\left(N_1 + \frac{\gamma_2}{4\gamma_1} - \frac{N_2 N_3}{1 - N_3^2}\right)\right]^2 \qquad (10.28)$$

$$z_1 z_1^* = N_3^2.$$

If we write

$$x_1 x_1^* - y_1 y_1^* = \varphi_1, \qquad (x_1 x_1^*)(y_1 y_1^*) = \varphi_2 \qquad (10.29)$$

we can find the wave normalized intensities dependence on the $\{N_1, N_2, N_3\}$, invariants as

$$x_1 x_1^* = \frac{2\varphi_2}{\sqrt{\varphi_1^2 + 4\varphi_2} - \varphi_1}, \qquad y_1 y_1^* = \frac{2\varphi_2}{\sqrt{\varphi_1^2 + 4\varphi_2} + \varphi_1}. \qquad (10.30)$$

10.3   Compensation solitons in non-stationary SBS process

We shall try to find solution in the form of hyperbolic secant, characteristic for solitons, using the hypothesis of 'isospectral evolution' [10.6–10.8, 10.12], which can be written in terms of the invariants $\{N_1, N_2, N_3\}$, in the form

$$N_2(\eta) = 0 \tag{10.31}$$

or as

$$I_S = \left[\frac{\pi\gamma_e}{n^2}\left(\frac{\rho}{\rho_0}\right)\right]^2 I_L \tag{10.32}$$

which leads to

$$(x_1 x_1^* - y_1 y_1^*)^2 = \left(\frac{1 - N_3^2}{N_3}\right)^2 (x_1 x_1^*)(y_1 y_1^*). \tag{10.33}$$

Using equations (10.30), one can derive the evolution equations for the normalized intensities $\{x_1 x_1^*, y_1 y_1^*\}$, on the characteristics $\{\Gamma_1, \Gamma_2\}$:

$$x_1 x_1^* = \frac{4\left(N_1 + \dfrac{\gamma_2}{4\gamma_1}\right)}{(1 + 3N_3^2)}, \qquad y_1 y_1^* = \frac{4N_3^2\left(N_1 + \dfrac{\gamma_2}{4\gamma_1}\right)}{(1 + 3N_3^2)}. \tag{10.34}$$

In this case, equations (10.25) lead to

$$\frac{\partial N_1}{\partial \eta} = -(4\gamma_1)N_1 - (\gamma_2), \qquad \frac{\partial N_3}{\partial \eta} = (4\gamma_1)N_3 + 2N_1 N_3 \tag{10.35}$$

which allows one to derive the implicit form

$$y_1 y_1^*(\eta) = N_3^2 x_1 x_1^*(\eta). \tag{10.36}$$

Equation (10.36) expresses the dependence of the normalized Stokes wave intensity on the normalized intensities of the optical pump and acoustic waves. It actually describes the usual amplification regime of the Stokes wave, when the optical pump (xx^*) is perturbed by the feedback induced by the other two fields.

If we calculate the second derivative of the second equation from (10.35) and use the first equation, the nonlinear Schrödinger-type equation is found [10.24]:

$$\frac{\partial^2 N_3^{-1}}{\partial \eta^2} = [4(N_1 + 3\gamma_1)^2 - 20\gamma_1^2 + 2\gamma_2]N_3^{-1}, \tag{10.37}$$

where

$$N_1(\eta) = c_1 e^{-4\gamma_1\eta} - \frac{\gamma_2}{4\gamma_1} \tag{10.38}$$

and c_1 is a constant of integration. In equations (10.28), imposing $N_3^2 \ll 1$, i.e. the small signal condition, one can obtain

$$N_1(\eta) + \frac{\gamma_2}{4\gamma_1} = \frac{1}{4}(x_1 x_1^* - y_1 y_1^*). \tag{10.39}$$

We can remark that the solution of equation (10.37) is a soliton (a hyperbolic secant, with continuous spectrum) only if the condition of the continuous spectrum is fulfilled:

$$4(N_1 + 3\gamma_1)^2 - 20\gamma_1^2 + 2\gamma_2 > 0. \tag{10.40}$$

Equations (10.40) and (10.39) lead to the following condition:

$$x_1 x_1^* - y_1 y_1^* > \frac{\gamma_2}{\gamma_1} - 12\gamma_1 + \sqrt{80\gamma_1^2 - 8\gamma_2} \tag{10.41}$$

which is precisely the necessary condition for the compensation soliton solution on the $\{\Gamma_1\}$ and $\{\Gamma_2\}$ characteristics. The sufficient condition for this soliton formation is given by the initial condition

$$N_3^{-1}(\eta)|_{\eta=\eta_0} = N_{30}^{-1}(\eta_0) \neq 0. \tag{10.42}$$

The soliton solution in the Stokes wave can be obtained from equation (10.37) as

$$y_1 y_1^*(\eta_0^\bullet) \cong \frac{1}{12} x'_{10}(\eta_0^\bullet) \, e^{-4\gamma_1(\eta_0^\bullet - \eta)}$$

$$\times \operatorname{ch}^{-2}\left[\ln\sqrt{\frac{3y_S}{x'_{10}(\eta_0^\bullet)}} + \frac{8\gamma_1^2 - \gamma_2}{2\gamma_1}(\eta_0^\bullet - \eta) + x'_{10}(\eta_0^\bullet)\frac{1 - e^{-4\eta(\eta_0^\bullet - \eta)}}{8\gamma_1}\right].$$

$$\tag{10.43}$$

For SBS in non-absorptive nonlinear media ($\alpha = 0$), equation (10.43) becomes

$$y_1 y_1^*(\eta_0'') = \left(\frac{x'_{10}(\eta_0'')}{12}\right) \operatorname{ch}^{-2}\left[\ln\sqrt{\frac{3y_S}{x'_{10}(\eta_0'')}} + x'_{10}\frac{\eta_0'' - \eta}{2} - \frac{A}{2\sigma}(\eta_0'' - \eta)\right]. \tag{10.44}$$

In the original coordinates, accounting the substitutions:

$$x_1 x_1^* \approx \frac{I_L}{I_0}, \qquad y_1 y_1^* \approx \frac{I_S}{I_0}, \qquad \alpha' = \frac{\alpha}{K}, \qquad \sigma_{1B} = g_B L'_B I_0,$$

$$\tag{10.45}$$

$$L_B = 2\lambda_{ac}, \qquad y_1 y_1^*(\eta_0'')|_{\eta_0''=\eta} = y_S \approx e^{-G}$$

the intensity of the Stokes field takes the form

$$I_S\left(t + \frac{n}{c}z_c\right) = \frac{1}{12} I_{L0}\left(t + \frac{n}{c}z_c\right)$$

$$\times \operatorname{ch}^{-2}\left\{\frac{1}{2}\ln\left[\frac{3I_{S0}}{I_{L0}\left(t + \frac{n}{c}z_c\right)}\right] + \frac{1}{L'_B} V_S(t, z_c)\left(t + \frac{n}{c}z_c\right)\right\}$$

$$\tag{10.46}$$

where $V_S(t, z_c)$ is the velocity of Stokes soliton along the characteristic line

$\varphi = \omega t + kz$ and can be written as

$$V_S(t, z_c) = \frac{\omega L'_B}{\delta} \left[g_B L'_B I_{L0}\left(t + \frac{n}{c}z_c\right) - 2\omega\tau \right]. \tag{10.47}$$

Equation (10.47) shows that the soliton velocity is proportional to the pump field, I_{L0} along the characteristic line. The time variations yield soliton accelerations and decelerations and the z-variations produce dispersion processes.

For SBS in absorptive nonlinear media ($\alpha \neq 0$), equation (10.43) can be written in the form [10.24]

$$I_S\left(t + \frac{n}{c}z_c\right) \cong \frac{1}{12} I_{L0}\left(t + \frac{n}{c}z_c\right) \exp\left[-8\gamma_1 \frac{\sigma\omega}{\delta}\left(t + \frac{n}{c}z_c\right)\right]$$

$$\times \text{ch}^{-2}\left[\ln\left(\sqrt{\frac{3I_{S0}}{I_{L0}\left(t + \frac{n}{c}z_c\right)}}\right) + \frac{8\gamma_1^2 - \gamma_2}{\gamma_1}\left(\frac{\sigma_{1B}\omega}{\delta}\right)\right.$$

$$\left. \times \left(t + \frac{n}{c}z_c\right) + I_{L0}\left(t + \frac{n}{c}z_c\right)\frac{1 - \exp\left[-8\gamma_1 \frac{\sigma\omega}{\delta}\left(t + \frac{n}{c}z_c\right)\right]}{8\gamma_1}\right]. \tag{10.48}$$

The necessary condition for SBS compensation solitons, (10.41), may be written as

$$x_1 x_1^*(\eta) - y_1 y_1^*(\eta) > \frac{2\omega}{\sigma_{1B}\Gamma_B} - \frac{1}{2}\frac{\alpha'}{\sigma_{1B}}\left(\delta\frac{\omega_L}{\omega}\right)\left\{1 - \sqrt{1 - \frac{1}{\delta}\frac{8\omega^2}{\Gamma_B\omega_L}}\right\}. \tag{10.49}$$

Taking into account that $\delta \approx 1$ and $8\omega^2 \ll \Gamma_B\omega_L$, equation (10.49) becomes

$$x_1 x_1^*(\eta) - y_1 y_1^*(\eta) > \frac{2\omega}{\sigma_{1B}\Gamma_B}. \tag{10.50}$$

The necessary condition for SBS compensation soliton solution (10.48) becomes

$$I_L - I_S > \frac{2\omega\tau(1 - \alpha')}{g_B L'_B} \tag{10.51}$$

which leads, with the isospectral condition (10.32), to

$$I_S = \left[\frac{\pi\gamma_e}{n^2}\left(\frac{\rho}{\rho_0}\right)\right]^2 I_L$$

and with the conditions of very low absorption and small Stokes intensity, to

$$I_L > \frac{2\omega\tau}{g_B L_B'} \left\{ 1 - \left[\frac{\pi\gamma_e}{n^2} \left(\frac{\rho}{\rho_0} \right) \right]^2 \right\}. \tag{10.52}$$

Thus, we can define the threshold intensity for the soliton existence as

$$I_{L_{thr}} = \frac{2\omega\tau}{g_B L_B'}. \tag{10.53}$$

Using a calibration length, L_B', equal to the characteristic length, $z_c = ct_L/n$, we can obtain the following condition for the soliton threshold intensity:

$$I_{L_{thr}} = \left(\frac{2n\omega}{g_B c} \right) \left(\frac{\tau}{t_L} \right). \tag{10.54}$$

In the particular case of CS_2 as the nonlinear SBS medium, at $\lambda_L = 1.06\,\mu m$, $n = 1.6$, $\omega = 2.4 \times 10^{10}$ Hz, $g_B = 0.06\,cm/MW$, $\tau = 6\,ns$, $t_L = 3\,ns$, this limit is: $I_{L_{thr}} = 80\,MW/cm^2$.

Looking at the soliton velocity from equation (10.47), one can remark that, for negligible losses, $I_{L0}[t + (n/c)z_c]$ is constant in time, inside the nonlinear medium and the soliton (envelope) velocity is [10.20–10.22]

$$V_s = \frac{\omega L_B'}{\delta} |g_B I_0 L_B' - 2\omega\tau| = \frac{\omega g_B L_B'^2}{\delta} (I_0 - I_{L_{thr}}). \tag{10.55}$$

Imposing that this velocity be smaller than the light velocity, one can derive an upper limit for the pump intensity as

$$I_0 < I_{L_{thr}} + \frac{\delta c}{\omega g_B L_B'^2}. \tag{10.56}$$

Thus, the smaller the pump pulse duration, the smaller the calibration length ($L_B' = ct_L/n$) and the larger the range of pump pulse intensities, in which one could observe SBS compensation solitons. For, CS_2 and the parameters considered above, $\delta \approx 1$ and the second term in (10.56) is $\Delta I \approx 5.76\,kW/cm^2$, a small quantity with respect to the first one. This means that the SBS compensation soliton could be observed in a very narrow range of pump pulse intensities. Assuming that it is possible to fix the pump intensity at $I_0 = 80\,MW/cm^2$, the soliton velocity results from equation (10.55) as $V_s \approx 2 \times 10^{10}$ cm/s.

The soliton pulse duration can be derived as

$$t_S = \frac{L_B'}{V_S} = \left| \frac{\delta}{\omega(g_B I_0 L_B' - 2\omega\tau)} \right|. \tag{10.57}$$

In CS_2 and with the previous parameters, the compensation soliton duration is $t_S = 3\,ns$, which is equal to the pump duration. Indeed, the soliton pulse formation is more favourable if the pump and Stokes pulse durations are smaller than the phonon lifetime, which justifies our initial choice of the

SBS transient model (the complete SBS equation set) to derive the soliton solution.

10.4 Topological solitons in SBS media with very low dispersion and absorption

The search for SBS solitons can start also from the normalized SBS equation in the phase space, equations (10.19), in the loss-less case ($\alpha' \approx 0$):

$$\hat{E}_L + \hat{E}_S \hat{E}_{ac} = c_1(\varphi_S)$$

$$\hat{E}_S - \hat{E}_L \hat{E}_{ac} = c_2(\varphi_L) \tag{10.58}$$

$$\frac{\partial \hat{E}_{ac}}{\partial \varphi} = -(2A)\hat{E}_{ac} + \sigma_{1B}\hat{E}_L \hat{E}_S$$

where $c_1(\varphi_S)$ is a prime integral on the characteristic (φ_S) and $c_2(\varphi_L)$ is a prime integral on the (φ_L) characteristic.

According to the theory of the algebraic invariants, which characterize the autonomous nonlinear differential equations, the two prime integrals occurring in equation (10.58) are proportional to the linear combination of the algebraic invariants. The simplest case is

$$c_1(\varphi_S) = x_0(\varphi_S), \qquad c_2(\varphi_L) = 0. \tag{10.59}$$

in this case, equations (10.58) become

$$\hat{E}_L(\varphi_L) = \hat{E}_{L0}(\varphi_S) - \hat{E}_S(\varphi_S)\hat{E}_{ac}(\varphi)$$

$$\hat{E}_S(\varphi_S) = \hat{E}_{ac}(\varphi)\hat{E}_L(\varphi_L) \tag{10.60}$$

$$\frac{\partial \hat{E}_{ac}(\varphi_L)}{\partial \varphi} = -(2A)\hat{E}_{ac}(\varphi) + \sigma_{1B}\hat{E}_L(\varphi_L)\hat{E}_S(\varphi_S).$$

Using some simple substitutions, equations (10.60) can be brought to the form

$$\hat{E}_L = \frac{\hat{E}_{L0}}{1 + \hat{E}_{ac}^2}$$

$$\hat{E}_S = \frac{\hat{E}_{L0}\hat{E}_{ac}}{1 + \hat{E}_{ac}^2} \tag{10.61}$$

$$\frac{\partial \hat{E}_{ac}}{\partial \varphi_L} = -(2A)\hat{E}_{ac} + \frac{\sigma_{1B}\hat{E}_{L0}^2}{(1 + \hat{E}_{ac}^2)^2}\hat{E}_{ac}$$

with

$$2A = 4\omega t \gg 0. \tag{10.62}$$

If the quality factor of the acoustic field, $\omega\tau$, is high, the system (10.61) accepts the following soliton solutions:

$$I_L(\varphi_L) = I_{L0} \frac{\exp\left[-2\int_{\varphi_{L0}}^{\varphi_L} g_B L_B' I_{L0}\, d\varphi_L + 8\omega\tau\varphi_L\right]}{4\mathrm{ch}^2\left[\int_{\varphi_{L0}}^{\varphi_L} g_B L_B' I_{L0}\, d\varphi_L - 4\omega\tau\varphi_L\right]} \tag{10.63}$$

$$I_S(\varphi_S) = \frac{I_{L0}(\varphi_S)}{4\mathrm{ch}^2\left[\int_{\varphi_{S0}}^{\varphi_S} g_B L_B' I_{L0}\, d\varphi_L - 4\omega\tau\varphi_S\right]} \tag{10.64}$$

$$\frac{\pi\gamma_e}{n^2}\frac{\rho}{\rho_0} = \exp\left[\int_{\varphi_0}^{\varphi} g_B L_B' I_{L0}\, d\varphi_L - 4\omega\tau\varphi\right]. \tag{10.65}$$

The solitons from equations (10.63–10.65) exist if

$$I_L > \frac{2\omega\tau}{g_B L_B'}, \qquad I_S = \left[\frac{\pi\gamma_e}{n^2}\frac{\Delta\rho}{\rho_0}\right]^2 I_L. \tag{10.66}$$

Thus, the conditions (10.66) are the existence conditions of the topological solitons, and other types of solitons in the Stokes (and pump) waves. The first condition is similar to that from (10.49), up to the additive small term. Using again the calibration length, $L_B' = z_c = ct_L/n$, we can obtain the following condition for the SBS topological soliton formation:

$$I_L > I_{L_{\mathrm{thr}}} = \left(\frac{2n\omega}{g_B c}\right)\left(\frac{\tau}{t_L}\right). \tag{10.67}$$

The equations (10.63–10.65) allow the calculation of the velocity and the time duration of the topological solitons as

$$V_{\text{soliton}} = 4\omega\tau c |g_B^e L_B' I_0 - 4\omega\tau|^{-1} = 2cI_{L_{\mathrm{thr}}}|I_0 - 2I_{L_{\mathrm{thr}}}| \tag{10.68}$$

and

$$\Delta t_S = \frac{L_B'}{V_{\text{soliton}}}. \tag{10.69}$$

The velocity of the Stokes soliton from equation (10.68) is valid in the case of the non-stationary isentropic SBS compression only, which imposes

$$\int_{\varphi_{L0}}^{\varphi_L} g_B L_B' I_{L0}\, d\varphi' > 4\omega\tau\varphi_L. \tag{10.70}$$

In the case of the isentropic expansion, defined by

$$\int_{\varphi_{L0}}^{\varphi_L} g_B L_B' \cdot I_{L0}\, d\varphi' < 4\omega\tau\varphi_L \tag{10.71}$$

one can notice from equations (10.63–10.65) and (10.67) that solitons can also exist, but the amplitudes of these soliton waves are close to that of the

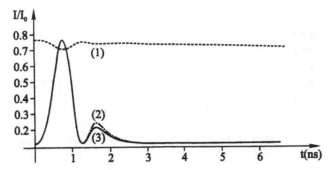

Figure 10.2. Time evolution of normalized pump intensity (1), Stokes intensity (compensation soliton) (2) and acoustic field intensity (3).

spontaneous scattered wave and their velocities are close to the phase velocity of the Stokes waves.

Imposing that this velocity be smaller than the light velocity, one can correct the lower limit for the pump intensity as

$$I_0 > 4I_{L_{thr}}. \tag{10.72}$$

In the particular case of CS_2 as the nonlinear SBS medium, using a laser at $\lambda_L = 1.06\,\mu m$ with pulses of $t_L = 3\,ns$, we can again take $n = 1.6$, $\omega = 2.4 \times 10^{10}\,Hz$, $g_B = 0.06\,cm/MW$, $\tau = 6\,ns$, and calculate $I_{0_{min}} = 320\,MW/cm^2$ from (10.72).

If we set $I_0 = 340\,MW/cm^2$, the velocity of the SBS topological soliton is $V_s = 2.6 \times 10^{10}\,cm/s$ and its duration is $\Delta t_s = 2.3\,ns$, which are comparable with those of the compensation soliton. We remark that the threshold intensity for the topological soliton formation is much higher than that of the compensation soliton and consequently more critical from the point of view of the material resistance to laser damage.

The pump, the Stokes and the acoustical wave intensities of the compensation and topological solitons are shown in figures 10.2 and 10.3, respectively. In these figures, the evolution curves of the pump (1) and the Stokes (2) wave intensities are actually the result of the projection of the two field profiles (from optical characteristics, $\varphi_{L,S}$, on acoustical field characteristic, φ, at $z = 0$).

In the case of the compensation soliton, a nonlinear collective interaction of the three waves takes place, when the dispersion is compensated by the SBS nonlinearity (described by a nonlinear Schrödinger-type equation). The existence condition of the compensation soliton, $g_B L'_B I_L > 2\omega\tau$, indicates the dominating role of the optical gain with respect to the acoustical one. In this case, a compensation soliton occurs in the acoustical wave, which yields the Stokes compensation soliton by a nonlinear feedback. The generation process of the topological soliton is different from the compensation one

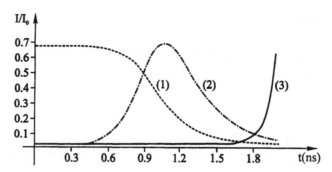

Figure 10.3. Time evolution of normalized pump intensity (1), Stokes intensity (topological soliton) (2) and acoustic field intensity (3).

(figure 10.3). For the topological soliton, a micro-compression of the non-linear propagation medium induces a nonlinear (almost total) absorption of the pump field. The consequence of the isentropic compression is the occurrence of a reflected optical (Stokes) field from the increased medium density. The acoustical gain must be higher in comparison with the optical gain, in order to allow such a process. The topological soliton appears at the threshold limit of the SBS process. It takes place at higher optical and acoustic intensities than required for the compensation soliton. The topological soliton duration is comparable with the compensation soliton duration.

10.5   Concluding remarks

In this chapter, the necessary and sufficient conditions for the temporal soliton generation in SBS were derived. We demonstrated that, in this nonlinear process, two types of solitons could occur. The usual compensation solitons arise from the compensation of the dispersion by the SBS nonlinearity, i.e. from the condition for obtaining solutions of a nonlinear Schrödinger-type equation with continuous spectrum. The topological solitons arise by particular linear conditions imposed on the algebraic invariants of the nonlinear equations, in the phase space.

The generation of these solitons differs by their mechanisms and by the threshold conditions. The compensation optical soliton is generated by the entire acoustic soliton, both of them being the result of the collective interactions of the light and acoustic waves. The topological solitons are generated by the fronts of the acoustic waves (i.e. by isentropic compressions and decompressions). The amplitudes, velocities and durations of the compensation and topological solitons are comparable. Both types of SBS solitons could be observed simultaneously in a space–time window, in the limits imposed by their parameters, which are quite critical for the usual experiments.

The model presented in this chapter can be used in the study of SBS solitons in communication systems [10.20-10.25]. Previous experimental results in the study of SBS solitons in mode-locked ring fibre lasers and in optical fibre communication, obtained by Montes *et al* [10.24, 10.25], seem to originate in the combined action of the two nonlinear processes: SBS and mode-locking.

References

10.1 Hasegawa A 1989 *Optical Solitons in Fibers* (Berlin: Springer)
10.2 Mollenauer L F, Gordon J P and Mamyshev P V 1997 in *Optical Fiber Telecommunications* vol IIIA (Lucent)
10.3 Agrawal G P 1995 *Nonlinear Fiber Optics* 2nd edition (San Diego: Academic Press)
10.4 Blow K J and Doran N J 1990 in *Nonlinear Waves in Solid State Physics* ed A D Boardman, M Bertolotti and T Twardowski (New York: Plenum)
10.5 Butcher P N and Cotter D 1990 *The Elements of Nonlinear Optics* (Cambridge: Cambridge University Press)
10.6 Calogero F and Degasperis A 1985 *Spectral Transform and Solitons* vol 1 (Amsterdam: North-Holland)
10.7 Ablowitz M J and Segur H 1981 *Solitons and the Inverse Scattering Transform* (Philadelphia: SIAM)
10.8 Novikov S, Manakov S V, Pitaevskii L P and Zacharov V E 1984 *Theory of solitons—the inverse scattering method* (Moskow: Akademia Nauk USSR) (in Russian)
10.9 Babin V, Mocofanescu A and Vlad V I 1996 *Ro. Repts. Phys.* **3-4** 299
10.10 Sagdeev R Z, Moiseev S S, Tur A V and Yanevskii V V 1988 Problems of the theory of strong turbulence and topological solitons in *Nonlinear Phenomena in plasma Physics and Hydrodynamics* (Moskow: Mir Publ)
10.11 Yariv A 1973 *Quantum Electronics and Nonlinear Optics* (New York: Wiley)
10.12 Babin V 1991 *Deterministic and Stochastic Models in Stimulated Brillouin Scattering* Doctoral Thesis, Institute of Atomic Physics, Bucharest
10.13 Kroll N M 1965 *J. Appl. Phys.* **36** 34
10.14 Maier M and Renner G 1971 *Opt. Commun.* **3** 301
10.15 Kliatkin V I 1986 *Averaging Problems and the Scattering Theory of Wave* (Moscow: Mir Publ) (in Russian)
10.16 Maier M 1968 *Phys. Rev.* **166** 113
10.17 Pohl D and Kaiser W 1970 *Phys. Rev. B* **1** 31
10.18 Hon D T 1980 *Opt. Lett.* **5** 516
10.19 Babin V and Vlad V I 1998 *Ro. Repts. Phys.* **50** 93
10.20 Liu Yushu, Li Hong and Zhang Shaowu 1999 *Chinese J. Quantum Electron.* **16** 276
10.21 Dennis M L, Carruthers T F, Kaechele W I, Jenkins R B, Kang J U and Duling I N 1999 *IEEE J. Photonics Tech. Lett.* **22** 478

10.22 Duling N 1999 *IEEE Photonics Technol. Lett.* **11** 478
10.23 Taranenko Y N and Kazovsky L G 1992 *IEEE Photonics Technol. Lett.* **4** 494
10.24 Babin V and Vlad V I 1999 *J. Optoelectr. Adv. Mat.* **1** 49
10.25 Montes C, Bahloul D, Bongrand I, Botineau J, Cheval G, Mamhoud A, Picholle E
         and Picozzi A 1999 *J. Opt. Soc. Am.* **B 16** 932
10.26 Montes C and Rubenchik A M 1992 *J. Opt. Soc. Am.* **B 9** 1857

Appendix

Averaging the Gaussian process describing the noise in SBS

The equation system which describes the SBS dynamics in the steady-state can be written as

$$\frac{dI_L}{d\eta} = -\alpha I_L - (1 - \varepsilon')g_B^e I_L I_S, \qquad \frac{dI_S}{d\eta} = -\alpha I_S + (1 - \varepsilon')g_B^e I_L I_S \quad \text{(A1.1)}$$

where ε' is a random variable which defines the noise as a Gaussian process and η is a characteristic coordinate. In this case, the variable ε' has the following properties

$$\langle \varepsilon'(\eta) \rangle = 0, \qquad \langle \varepsilon'(\eta)\varepsilon'(\eta') \rangle = 2\sigma^2 \delta(\eta - \eta')g_B^e I_0, \quad \text{(A1.2)}$$

where $\langle \cdots \rangle$ is the representation of the average of the Gaussian process $\varepsilon(\eta)$, σ is the dispersion of this Gaussian noise in the SBS, and $\delta(\eta - \eta')$ is the Dirac function.

We use the formalism of the statistic Liouville equation, described in [3.19–3.22] for a function $\psi_\eta(I_L, I_S)$ defined in the space $\{I_L, I_S\}$. The Liouville equation has the form

$$\frac{\partial \psi_\eta(I_L, I_S)}{\partial \eta} = \frac{\partial}{\partial I_L}\left[\frac{dI_L}{d\eta}\psi_\eta(I_L, I_S)\right] - \frac{\partial}{\partial I_S}\left[\frac{dI_S}{d\eta}\psi_\eta(I_L, I_S)\right]. \quad \text{(A1.3)}$$

Introducing the explicit form of the variable $\varepsilon'(\eta)$ in the statistic Liouville equation, we obtain from relations (A1.1) and (A1.3):

$$\frac{\partial \psi_\eta(I_L, I_S)}{\partial \eta} = \hat{U}_1 \psi_\eta(I_L, I_S) - \varepsilon'(\eta)g_B^e \hat{U}_2 \psi_\eta(I_L, I_S) \quad \text{(A1.4)}$$

where the operators $\hat{U}_1$ and $\hat{U}_2$ have the form

$$\hat{U}_1 = \frac{\partial}{\partial I_L}(\alpha I_L + g_B^e I_L I_S) + \frac{\partial}{\partial I_S}(\alpha I_S - g_B^e I_L I_S)$$

$$\hat{U}_2 = \frac{\partial}{\partial I_L}(I_L I_S) - \frac{\partial}{\partial I_S}(I_L I_S). \quad \text{(A1.5)}$$

191

Averaging the Liouville equation (A1.4) on the Gaussian process $\varepsilon'(\eta)$ and using the method developed in [3.20–3.21], we obtain the Fokker–Planck–Kolmogorov equation

$$\frac{\partial P_\eta(I_L, I_S)}{\partial \eta} = \hat{U}_1 P_\eta(I_L, I_S) - \sigma^2 g_B^e \hat{U}_2^2 P_\eta(I_L, I_S) \tag{A1.6}$$

where

$$P_\eta(I_L, I_S) = \langle \psi_\eta(I_L, I_S) \rangle \tag{A1.7}$$

is a repartition function.

Using the initial conditions for the SBS process, we obtain the relation

$$\int_0^1 P_\eta(I_L, I_S) \, dI_L \, dI_S = I_0^2 \tag{A1.7'}$$

where I_0 is an arbitrary normalization constant.

We define the average values for I_L and I_S as

$$\langle I_L \rangle = \frac{1}{I_0^2} \int_0^1 \int I_L P_\eta(I_L, I_S) \, dI_L \, dI_S$$

$$\langle I_S \rangle = \frac{1}{I_0^2} \int_0^1 \int I_S P_\eta(I_L, I_S) \, dI_L \, dI_S. \tag{A1.8}$$

In the papers [3.20–3.21], numerical calculation (or analytical treatment) are made for the repartition function P_η which contains almost all information on the process.

Applying the averaging operators defined in equations (A1.8) directly on the system (A1.1), we obtain a deterministic system with average variables of the Gaussian process, $\varepsilon'(\eta)$

$$\frac{\partial \langle I_L \rangle}{\partial \eta} = \frac{1}{I_0^2} \int_0^1 \int I_L \hat{U}_1 P_\eta(I_L, I_S) \, dI_L \, dI_S$$

$$- \sigma^2 g_B^{e2} \frac{1}{I_0^2} \int_0^1 \int I_L \hat{U}_2^2 P_\eta(I_L, I_S) \, dI_L \, dI_S$$

$$\frac{\partial \langle I_S \rangle}{\partial \eta} = \frac{1}{I_0^2} \int_0^1 \int I_S \hat{U}_1 P_\eta(I_L, I_S) \, dI_L \, dI_S \tag{A1.9}$$

$$- \sigma^2 g_B^{e2} \frac{1}{I_0^2} \int_0^1 \int I_S \hat{U}_2^2 P_\eta(I_L, I_S) \, dI_L \, dI_S.$$

The integrals from the equation system (A1.9) are calculated by parts, introducing the explicit form of the operators $\hat{U}_{1,2}$ defined in equations (A1.5).

The initial conditions of the SBS process introduce some limitations for the repartition function $P_\eta(I_L, I_S)$.

$$P_\eta(I_L, I_S)|_{I_S = I_0} = 0, \qquad P_\eta(I_L, I_S)|_{I_L = I_{L0}} = 0$$

$$\left.\frac{\partial P_\eta(I_L, I_S)}{\partial I_L}\right|_{I_L = I_{L0}} = 0, \qquad \left.\frac{\partial P_\eta(I_L, I_S)}{\partial I_S}\right|_{I_S = I_{L0}} = 0. \tag{A1.10}$$

The repartition function $P_\eta(I_L, I_S)$ has the physical significance of the density probability that I_L and I_S are in a certain state.

Finally, with the condition (A1.10), the equation system (A1.9) has the deterministic form

$$\frac{\partial \langle I_L \rangle}{\partial \eta} = -\alpha \langle I_L \rangle - g_B^e \langle I_L I_S \rangle - \sigma^2 g_B^{e2} [\langle I_L I_S^2 \rangle - \langle I_L^2 I_S \rangle]$$

$$\frac{\partial \langle I_S \rangle}{\partial \eta} = -\alpha \langle I_S \rangle + g_B^e \langle I_L I_S \rangle - \sigma^2 g_B^{e2} [\langle I_L^2 I_S \rangle - \langle I_L I_S^2 \rangle] \tag{A1.11}$$

which leads to the evolution equations (3.45) of the mean values of the pump and Stokes intensities.

Index